Alf Staudach

Sectional Fetal Anatomy in Ultrasound

Forewords by
W. Thiel, M. Hansmann and J. Hobbins

Translated by Terry Telger

Reviewed by Bernd K. Wittmann

With 247 Figures

Springer-Verlag
Berlin Heidelberg New York London Paris Tokyo

Prof. Dr. Alf Staudach
Landesfrauenklinik, Landeskrankenanstalten Salzburg
Müllner Hauptstraße 48, 5020 Salzburg, Austria

Translator
Terry Telger
6112 Waco Way, Ft. Worth, TX 76133, USA

Revisor
Bernd K. Wittmann, M.D.
Department of Diagnostic Ultrasound, Grace Hospital
4490 Oak Street, Vancouver BC V6H 3V5, Canada

ISBN-13:978-3-642-72918-8 **e-ISBN-13:978-3-642-72916-4**
DOI: 10.1007/978-3-642-72916-4

Library of Congress Cataloging-in-Publication Data. Staudach, A. (Alfons) [Fetale Anatomie im Ultraschall. English] Sectional fetal anatomy in ultrasound / Alf Staudach : forewords by W. Thiel, M. Hansmann, and J. Hobbins : translated by Terry Telger : reviewed by Bernd K. Wittmann.
p. cm. Translation of: Fetale Anatomie im Ultraschall.
Bibliography: p. Includes index. ISBN 0-387-18213-6 (U.S.) 1. Fetus–Anatomy. 2. Fetus–Ultrasonic imaging. I. Title.
RG605.S7313 1987 611'.013–dc19 87-22481

Softcover reprint of the hardcover 1st edition 1987

2121/3140-543210

Foreword

Alfons Staudach has been a long-time member of the Anatomic Institute of Karl Franzens University in Graz, where he has devoted particular attention to the deeper understanding, appreciation and visualizion of gross anatomic details.
In this work the author has achieved correspondence between sonograms and anatomic sections with a consistency and persuasiveness unequaled in all the previous literature on diagnostic ultrasound. The various planes of section and their characteristic features, and indeed the entire format of the text, are designed to provide even the less experienced sonographer with a valuable basis for conducting his examinations. The more experienced reader will find essential information on topographic relations and organ development that is not available in any other work dealing with fetal anatomy.
I am certain that my high estimation of this volume will prove justified, and that it will provice its readers with a useful and stimulating resource.

Univ.-Prof. Dr. Walter Thiel
(Chairman of the Anatomic Institute
of the University of Graz)

Foreword

Anyone setting this book down after an initial perusal must wonder why such a reference was not available ten years ago. The meticulous and fascinating juxtaposition of gross anatomic sections with sonograms, together with explanatory drawings and many practical guidelines, should enable even the novice accurately to identify details and interpret sonographic findings with precision. The time may finally be past when the haphazard labeling of sonographic features places ultrasound in danger of being dismissed as unscientific. At the same time, it must be acknowledged that a text of this kind probably owes its existence to technical advances in instrumentation. Today we can work with image resolutions that could not have been dreamed of years ago. Yet comparison of the anatomic sections with the sonograms shows that even more progress in this area can be achieved. In that regard, the present book may be considered a basis for future developments. No other author could be as predestined for this subject as Dr. Alf Staudach, who combines years of experience in anatomy with daily critical evaluation and interpretation of sonograms in obstetrics and gynecology. Added to this is his talent with pen and drawing pencil.
The reader will benefit greatly from the extensive practical experience of the author, whose latest work is sure to become a standard in prenatal ultrasonography.

Prof. Dr. M. Hansmann

Foreword

Ten years ago, there were only a handful of textbooks available in the English language specifically devoted to ultrasound in obstetrics and gynecology. Now, however, the number of texts of this kind emerging each year exceeds the total number of texts published prior to 1976. Also, since the scope of obstetrical ultrasound has grown exponentially, texts will soon be focused on only one facet of obstetrical and gynecological scanning alone, such as infertility, first trimester diagnosis, fetal anomalies, etc. Because of this predictable evolution, this may be one of the last textbooks of its kind, *and* one of the very best.

The sonographer and sonologist can no longer simply determine the position of the fetus and placenta, take a few measurements, and send the patient home. We live in an era in which our colleagues and patients demand that we provide them with all the information that can possibly be gathered from an ultrasound examination. Occasionally, the expectations of our patients and their legal representatives exceed what a mortal diagnostician can provide. Therefore, our best protection is identical to what is best for our patients: A consistent, comprehensive examination in which every area of the uterus (or pelvis) is evaluated systematically.

In Europe it is commonplace to scan all patients routinely. In the United States this is not uniformly accepted. The "hard data" from randomized clinical trials are beginning to show trends that suggest the benefit of routine screening with ultrasound (although changes in perinatal mortality must await studies containing many thousands of patients).

In the meantime, we must follow our own instincts and conscience in utilizing this powerful diagnostic tool. In helping to do this the author has taken on all aspects of obstetrical and gynecological ultrasound. Because the format is manageable and the author is concise and consistent in his approach to the subject, he has managed to create an excellent resource for the practicing obstetrician/gynecologist.

John Hobbins
New Haven

Table of Contents

1 Introduction

In the last 25 years sonography has become an integral part of obstetric diagnosis. No other modality permits such direct visual inspection of the fetus, analysis of its anatomic structures, and evaluation of its behavior.
The improved resolution of modern equipment permits an increasingly detailed identification of fetal anatomic structures. However, prenatal ultrasound differs from pediatric and adult ultrasound in several important respects:

1. In pediatric and adult ultrasound, the examiner is able to relate surface anatomy directly to intracorporeal anatomy. In obstetric ultrasound this orientation is lacking, since the transducer is not directly in contact with the fetal surface.
   It is therefore important to develop a standard examination protocol. The examiner must first identify the presentation, position and attitude of the fetus in the uterus and at the same time follow fetal movements. Only then is it possible to select the appropriate anatomic reference planes and analyze the individual organs.
2. While textbooks of anatomy contain a large number of standardized anatomic sections, the present author is not aware of any work that presents the anatomic reference planes which have to be obtained in order to perform accurate biometry and anatomic screening of the fetus.
3. Organ development during the embryonic period has been extensively analyzed in embryologic texts and described in detail. The same is true of the neonatal period. However, the interpretation of fetal sonograms is difficult because essential information on the changes in fetal organ morphology is lacking, leaving an inadequate knowledge basis from which to differentiate normal development and normal variation from pathological changes. The pediatric principle that "a child is not a small adult" may be extended by noting that a fetus is not simply a small child.
4. A full comprehension of the ultrasound anatomy of the fetus can be gained only by comparing sonographic morphology with corresponding anatomic sections. In the past, for such comparisons, which were few and far between, formalin-fixed sections from adult cadavers were used (Johnson and Rumack 1980; Hadlock et al. 1981; Grant et al. 1981). Recently this circumstance has led to misinterpretation of certain details of fetal anatomy (Campbell and Thoms 1982; Chinn et al. 1982) and to the outright mislabeling of some structures (Jeanty et al. 1984).

In view of these factors, the author prepared frozen sections corresponding to the reference planes used in obstetric sonography for biometry and organ studies. With the knowledge gained from this work an examination protocol was developed which was standardized according to anatomic criteria. Where either improved image quality or better anatomic understanding led to the sonographic detection of new structures that could not be readily identified, new frozen sections of the region were prepared in an effort to clarify the finding. In the six-year period from 1980 to 1985, all fetal ultrasound examinations were performed in accordance with the examination protocol derived from the above experience, and the value of the protocol as a tool for the evaluation of the fetus was analyzed.

2 Basic Principles

2.1 Frozen Section Technique; Photographic Documentation

To gain a deeper insight into fetal anatomy, we prepared fetal anatomic sections analogous to the sections obtained on ultrasound scans. Our objectives were as follows:

1. To provide information on anatomic areas that are not common knowledge to sonologists.
2. To establish the true nature of structures seen on fetal sonograms (does a linear echo on a transverse scan through the brain really represent the third ventricle?). It was our goal to identify the anatomic nature of such structures unambiguously through in-vitro experimentation (water-bath scanning, marking of the structures in question, identification on anatomic sections).
3. To define precisely the scan planes necessary for fetal anatomic screening, and to define anatomically the reference planes necessary for accurate fetal biometry.
4. To analyze sections from fetuses of varying gestational ages and thus elucidate the dimensional dynamics of organs and structures that undergo major changes in size during fetal growth (determination of ventricular size as a function of gestational age).

There were basically three options available in terms of specimen preparation for these studies:

1. Traditional formalin fixation and sectioning.
2. Frozen sections.
3. Plastination.

Of these options, we felt that frozen sections were most appropriate for the demonstration of fetal anatomy. At the time our studies began, the technique of plastination (Klemstein 1981) was not perfected, and the necessary equipment was not readily obtainable.
Formalin fixation was ruled out because it produces shrinkage effects that prevent an accurate biometric correlation with sonoanatomy, especially in the brain (Fig. 2.1 a, b) (Bahr et al. 1957; Kushida 1962; Boonstra et al. 1984; Tsukasa et al. 1984).
Frozen section studies were performed on a total of 122 embryos and fetuses during the period from 1/1/76 to 12/31/85. Sections obtained from 86 fetuses between 11

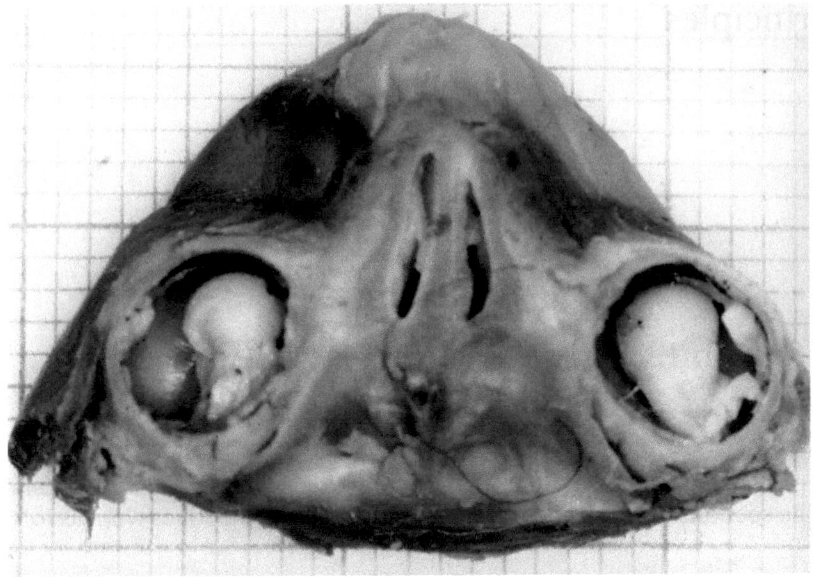

Fig. 2.1. a Formalin preparation of the fetal orbital region. The formalin fixation has led to marked shrinkage of the globes

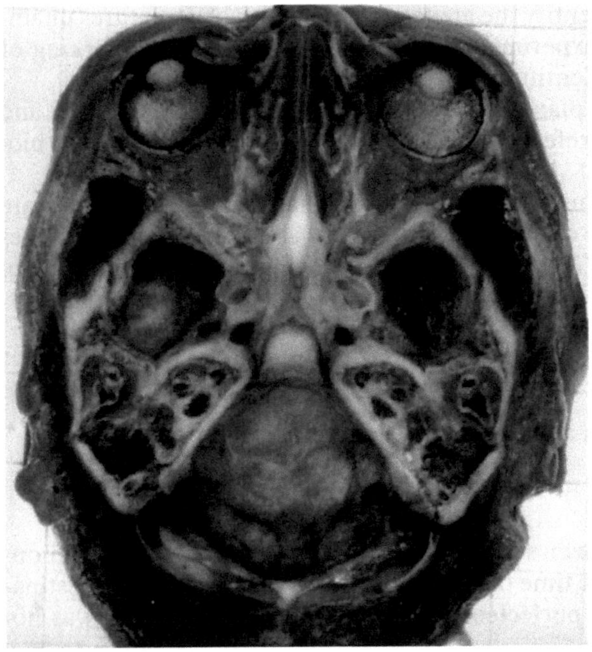

Fig. 2.1. b Frozen section through a fetal skull at the level of the eyes. The ocular dimensions show no shrinkage effects

and 24 weeks' gestation were used for the present investigation. All the fetuses were products of spontaneous abortions, and all sectional studies were performed on fetuses whose gestational age had been established with high confidence by the menstrual history (regular cycle, LMP) and by measurement of the crown-rump length in the first trimester. All the fetuses were stillborn, and each was carefully examined to exclude vitality before any manipulations were carried out. In all cases the parents gave explicit consent to the use of the bodies for scientific purposes. Experimental water-bath scans were obtained in 12 cases in an attempt to clarify sonoanatomic features. We used distilled water heated to 100 °C to eliminate artifacts from gas bubbles. The results of these investigations were unsatisfactory. The quality of organ imaging in vitro could not be compared to the quality of in vivo examinations. In most cases the blood vessels were poorly visualized due to intravascular coagulation, and this greatly compounded the problem of intrafetal orientation. Images of acceptable quality were obtained in the region of the central nervous system. In cases where structures were reproducibly observed in intrauterine examinations but could not be readily identified (these were mostly intracranial), a needle was inserted into the structure under ultrasound guidance, and a small depot of methylene blue was injected to mark the site so that it could later be identified in the anatomic section.

Freezing was induced in a large cryostat chamber at −20 °C. Because supine or prone positioning caused the nonturgid trunk to become misshapen before hardening, the fetuses were positioned vertically. Comparative tests on fetuses fixed in various rotational positions (through 180°) showed no significant shifts in intrafetal anatomy with position changes during freezing.

In cases where sectioning on standard planes was planned, we marked the desired planes of section using known landmarks on the body surface (Figs. 2.2 and 2.3). The planes were marked either with twine loops, which left visible surface impressions when applied during initial freezing, or with thin paper tapes, which became adherent to the body surface during the freezing process. Because frozen-section microtomes were unavailable for preparing sections of the desired size, we used commercially available disk-type slicers. The markings on the fetal body surface could be easily followed during sectioning. Symmetry and thickness were carefully checked on th first sections obtained. With practice, we were able to prepare intact sections from 2 mm to 5 mm in thickness. Sections were kept frozen for periods of 48 h to 4 weeks before they were studied. Freeze-drying effects were evident if the sections were stored for prolonged periods.

To define reference planes in three dimensions for measurement of the biparietal diameter (BPD), occipito-frontal diameter (OFD), and for trunk biometry, we determined the corresponding, diametrically opposite points on the body surface spaced a maximum distance apart, and measured and marked them before freezing was initiated. We marked the points with needles passed completely through the corresponding body section (Fig. 2.4). A circumferential mark was placed through the needle exit sites so that the section could be made accurately in the plane defined by the two intersecting lines. Sections were then cut parallel to that plane until the marked area was reached (Fig. 2.4).

To obtain three-dimensional anatomic information for specific areas on the surface of the sections, we either prepared serial sections, or we selected a plane well defined by its surface anatomy and sectioned along that plane at a 90° angle to provide a specimen useful for three-dimensional orientation. Figure 2.5 a, b shows two views of a midsagittal section through a fetal head. The brain structures are

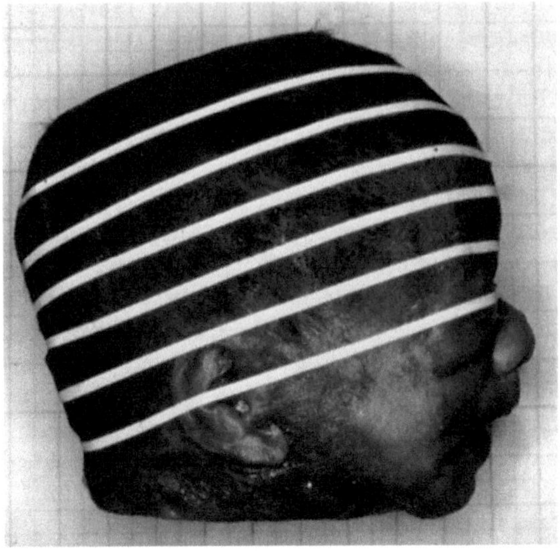

Fig. 2.2. Example of the surface markings used to guide sectioning of the fetal cranium

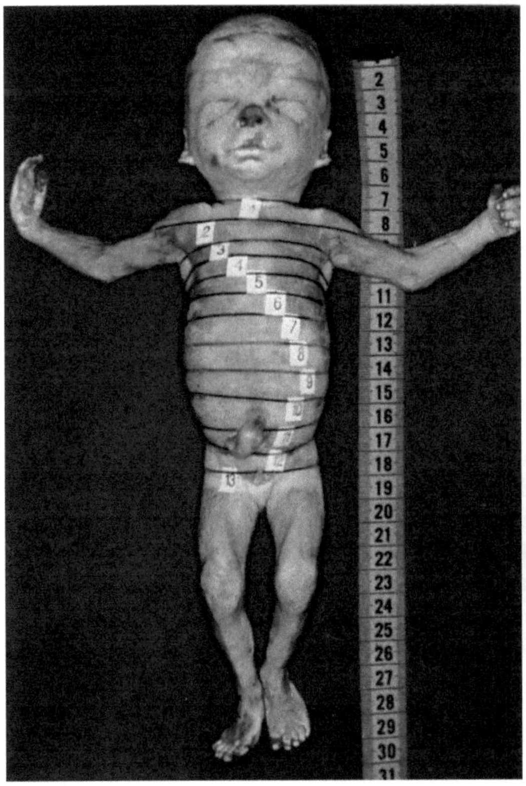

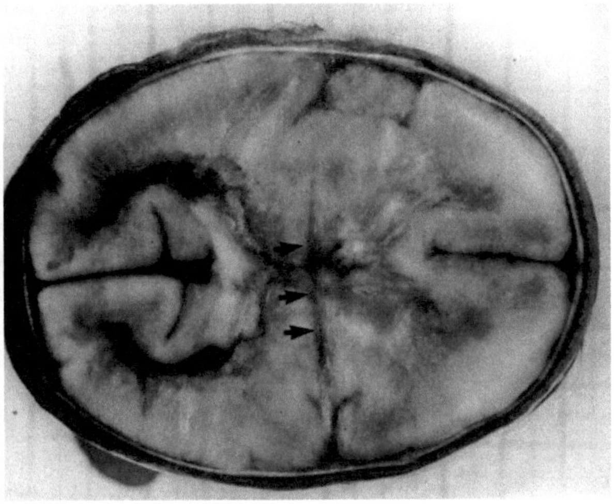

Fig. 2.3. (left) Markings on the fetal trunk to guide transverse sectioning of the thorax and abdomen

Fig. 2.4. (right) Transverse section through a fetal brain at 23 weeks. The *arrows* mark the track left by the needle that was inserted to establish the reference plane for the BPD. The needle track extends through both thalami

easily identified on the cut surface of the sagittal section (cerebellum, lamina tecti, pons). The overhead view in Fig. 2.5 b (transverse section 90° to the sagittal plane) demonstrates the left lateral ventricle. Each time we prepared a section of this kind, we attempted to define the anatomic structures that would be expected to appear in the next section before that section was cut, and we monitored the accuracy of the prospective orientation based on the structures actually exposed. This three-dimensional mental approach to the visualization of anatomic structures forms an important basis for the early detection of fetal anomalies by ultrasound.

The frictional heat of sectioning caused some thawing of surface areas, which refroze into "beads" a few seconds after passage of the blade because of the low temperature of the surrounding tissues. To prevent this phenomenon, which makes the specimen difficult to photograph, we removed liquid drops with an absorbent linen cloth immediately after passage of the blade.

For photographic documentation we used the Olympus OM-2n, OM-2 Spot, and OM-4 cameras equipped with an Olympus Zuiko 50-mm f 3.5 macro lens. An Olympus T-10 ring flash was used for flash exposures. We used Kodak Ektachrome ED-200 film (200 ASA/24 DIN) for color prints, and after some experimentation we selected Kodak Panatomic-X film (32 ASA/16 DIN) for black-and-white photos. The film sensitivity of the camera was adjusted to the type of film used. The aperture setting was from 11 to 22, depending on the range, and exposure time was controlled automatically by the camera. For flash pictures the camera was set to Full Automatic, a mode in which the light output of the flash is controlled automatically by the objective lens. Because the surface of frozen sections is always covered by tiny ice crystals despite cleaning, troublesome reflections are created which make the photographs more difficult to interpret (Fig. 2.6). To correct this, we attached a polarizing filter over the ring flash and adjusted the film sensitivity of the camera accordingly (recommended +1.5 stops). This resulted in a slight softening of image contrasts.

To satisfy biometric requirements, the sections were placed on millimeter grid paper before they were photographed. Of the various colors and types of paper available, transparent paper with gray millimeter grid lines placed over a white background proved most acceptable. The advantage of the transparent paper is its resistance to fluid uptake. The longer it took to photograph a section, and the thinner the section was, the more the specimen tended to thaw. To avoid deformation, the sections were returned to the freezer immediately after they were photographed. It was decided to keep the sections on the grid paper during freezer storage; the bottom surface of the section did not freeze solidly to the paper owing to its low absorbency, and this gave us the option of photographing that surface as well if additional documentation was required.

More by coincidence than by scientific analysis, we found that the surfaces of frozen sections stored for prolonged periods exhibited freeze–drying effects that brought out anatomic details without causing significant shrinkage or deformation. Both pictures in Fig. 2.7 a, b are of the same specimen–one photographed immediately after sectioning (a) and the other after 4 weeks' freezer storage (b). Because this process was reproducible and consistently enhanced surface details, regardless of the body region involved, we subsequently made it a practice to wait 3–4 weeks before photographing the specimens.

Our experience indicates that this period is optimum in terms of photodocumentation. If sections are stored for longer periods, they become dessicated to the point where shrinkage and structural loss become a problem.

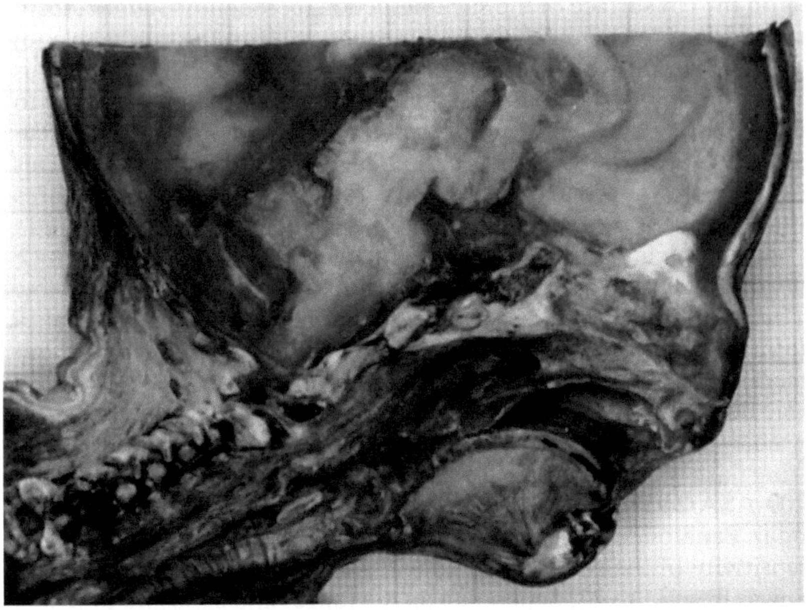

Fig. 2.5. a Midsagittal section through a fetal skull. This view is helpful in planning the placement of the transverse sections

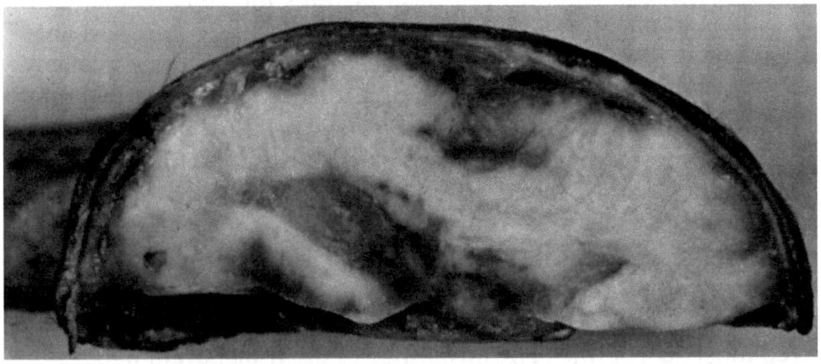

Fig. 2.5. b Surface of the transverse section viewed from above. This section passes through the left lateral ventricle

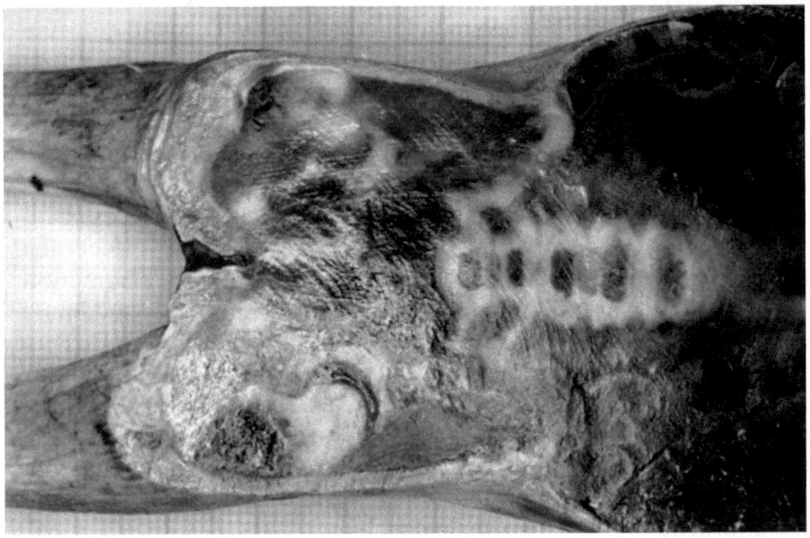

Fig. 2.6. Typical example of poor photodocumentation. Reflections from ice crystals obscure anatomic details

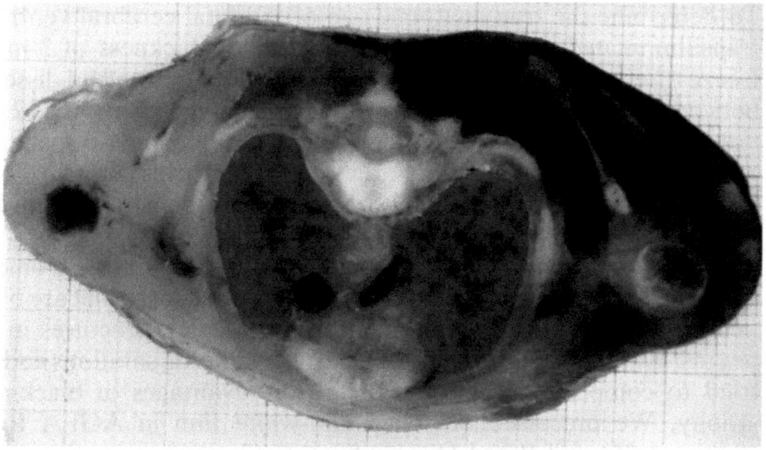

Fig. 2.7. a Transverse section through a fetal thorax. Photo was taken immediately after the section was performed

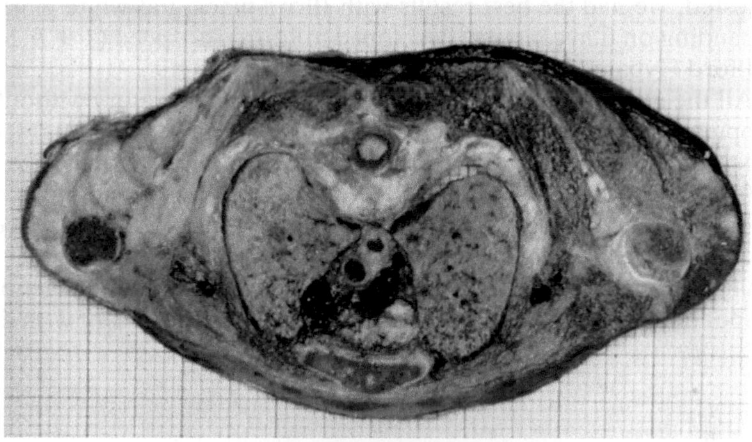

Fig. 2.7. b Photo of the same section taken four weeks later. Surface details have been greatly enhanced by freeze drying

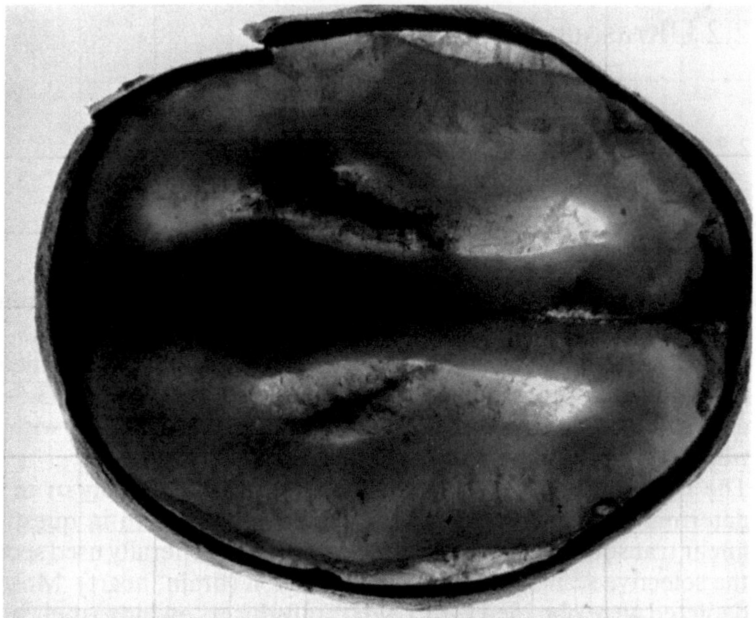

Fig. 2.8. Transverse section through a fetal brain at 17 weeks. The transillumination effect demarcates the boundary between the ventricles and brain mantle. The back-lighted choroid plexus is visible at the center of the ventricles and in the occipital horns

To determine the true anatomic size of the fetal cerebral ventricular system, we transilluminated sections having a maximum thickness of 3 mm that had been freeze dried for 3–4 weeks. This enabled us to visualize and dissect the boundaries between the fluid-filled ventricles and surrounding brain (Fig. 2.8). Direct dissection of the boundary between the brain and subarachnoid space was unnecessary. After dissection of the central area of the ventricles, pressure on the ice crystals with small lancets was sufficient to dislodge the cerebrospinal fluid crystals from the ventricular wall.

The analysis of anatomic details is more easily accomplished on color photos than on black-and-white because of the many visual nuances that are obtained. This particularly applies to the differentiation of vascular structures in parenchymatous organs. By experimenting with different developing solutions and paper grades, we tried to compensate for some of the disadvantages of black-and-white photography. We processed the black-and-white film in AGFA Rodinal developer (6 min, 1:25) and Ilford Hypam fixative (2 min, 1+4). Of the different papers tested, we had the best results with Ilford black-and-white glossy grade 0–2 (depending on film contrast) developed in Ilford Ilfospeed (1 + 9.1 min) and fixed in Ilford Hypam (1 + 9.1 min).

All the illustrations were prepared in our own laboratory without the use of developing machines. There were continuous alternation and interaction between frozen section studies and ultrasound examinations. Experience gained from the frozen sections was applied to the interpretation of sonograms, and questionable sonographic findings were resolved by the preparation of new sections. To convey our experience to the reader in the clearest manner possible, we arranged the illustrations in a way that would juxtapose frozen anatomic sections with the corresponding sonograms, supplementing this where necessary with drawings.

2.2 Ultrasound Examinations

2.2.1 Materials and Methods

Ultrasound Equipment

Several different instruments were used for ultrasound examinations during the period of the study: the Aloka Echocamera SSD-256 with a 3.5- and 5-MHz linear transducer, the Kranzbuhler 81/50 unit with a 3.5-MHz linear transducer and 3.5-MHz sector transducer, the Squibb Ultramark 4 with a 3.5-MHz linear transducer and 5-MHz sector transducer, and the Acuson 128 with a 3.5 MHz linear transducer and 3.5-MHz sector transducer.

The choice of a linear or sector transducer, individually or in combination, was determined by the nature of the diagnostic problem in question. We preferred linear transducers for initial survey views, and generally used sector transducers for the selective scanning of a particular region (brain, heart). Most of the sonograms depicted were obtained with 5-MHz transducers owing to their superior resolution.

Photographic Documentation

All sonographic images were recorded on 35-mm film, using either built-in photomonitors or separate cameras coupled to the video output of the instrument. We used the Olympus OM-10, OM-2n and OM-4 cameras (through-lens metering) with an Olympus Zuiko 50-mm f 3.5 macro lens. Exposures were made on Ilford HP-5 black-and-white film (400 ASA/27 DIN). The camera was set to 400 ASA and an aperture of f 11; shutter speed was controlled automatically. The film was developed in AGFA Rodinal (6 min, 1:25) and fixed in Ilford Hypam (2 min, 1 + 4). Prints were made on Ilford glossy paper grade 0–2 (depending on film contrast), which was developed in Ilford Ilfospeed (1 + 9.1 min) and fixed in Ilford Hypam (1 + 9.1 min).

Clinical Application

The ultrasound examinations were conducted under standardized conditions (see p. 23 Examination Procedure) and were performed primarily by the author. Technologists do not generally perform ultrasound examinations in our department. A total of 36,405 examinations were performed on obstetric patients at Salzburg Women's Clinic during the period from January 1, 1980, to December 31, 1985. Of that total, 3218 women were examined by the author for the primary purpose of evaluating the integrity of fetal anatomy. This population was divided prospectively into four main groups based on the indication for referral:

Group 1 consisted of women undergoing baseline routine examinations. There was no preliminary evidence of abnormality in these patients, and none were referred for the exclusion of fetal anomalies on the basis of known risk factors. The question if the patients had undergone ultrasound examinations prior to referral could not be answered in this group due to a lack of specific information.

Group 2 consisted of women who were referred for the exclusion of fetal anomalies because of their histories (family history of developmental anomalies, malformations in previous offspring, exposure to potential teratogenic agents in the first trimester).

Group 3 consisted of women in whom preliminary ultrasound examinations had disclosed indirect warning signs (abnormal amniotic fluid volume, suspicion of intrauterine growth retardation (IUGR), discrepancies in biometric parameters), and who were candidates for reexamination and selective evaluation of fetal anatomy.

Group 4 consisted of cases in which the previous examiner had expressed suspicion of a fetal anomaly.

Evaluation and Follow-Up

In every case the result of the examination was documented, and an opinion was rendered: normal sonoanatomy or a detectable anomaly. In patients referred with a suspected fetal malformation, this impression was confirmed or refuted by several examinations performed together with specialists from other areas (pediatrics, pediatric surgery, pediatric neurology, neurology, radiology), and an effort was made to arrive at a consensus opinion.

To check the accuracy of the diagnosis in cases where sonographic findings had been declared normal, babies delivered at the Salzburg Women's Clinic were evaluated at the time of delivery by the attending obstetrician and, while in hospital, underwent two pediatric examinations and one orthopedic examination. In cases where suspicion of an anomaly had been expressed, the babies were transferred directly to the neonatal unit and evaluated using appropriate diagnostic procedures.

Postmortem examinations were performed in all spontaneous abortions and perinatal deaths (Pathology Institute of the Landeskrankenanstalten Salzburg, Prof. Thurner). Prior to the examination, the pathologist was provided with the presumptive diagnosis in cases where suspicion of an anomaly had been expressed on the basis of sonographic findings.

In cases where the delivery, abortion, or perinatal death did not occur in our department, we contacted the departments involved and obtained the information necessary to establish a definitive diagnosis. Follow-up information was also obtained for babies delivered in other departments where sonographic findings had been normal.

We compared prenatal diagnostic findings with the definitive information after delivery and evaluated them by individual case analyses so that we could determine the accuracy and limitations of ultrasound in the prenatal diagnosis of fetal anomalies.

2.2.2 Results

Study Population

In Table 2.1 the 3218 women are grouped by indications for referral. Eighty percent of the women (2569) had no anamnestic risk and were referred for a basic routine examination. Patients with an anamnestic risk, indirect warning signs, or a suspected fetal anomaly comprised about 6%–7% each of the total population.

Trends

When we analyse the study period by individual years, we observe a progressive rise in patient numbers in all categories (Table 2.2). This is most pronounced in the anamnestic risk group, and especially in women with a familial risk. The total number of referrals nearly quadrupled in the period from 1980 to 1985. This trend is due

Table 2.1. Indications for referral of the 3218 women in the study population

	[*n*]	[*n*]	[%]
1. Routine examination		2569	80
Familial risk	152		
Exogenous insult	60		
2. Anamnestic risk		212	6.5
Hydramnios	113		
Oligohydramnios	57		
IUGR	42		
3. Indirect warning signs		212	6.5
4. Suspected fetal anomaly		225	7.0
Total		3218	100

Table 2.2. Patient distribution by referral groups on a yearly basis between 1980 and 1985

	1980	1981	1982	1983	1984	1985	Total
1. Routine examination	212	314	421	478	512	632	2569
Familial risk	4	8	6	12	27	95	152
Exogenous insult	3	4	12	7	9	25	60
2. Anamnestic risk	7	12	18	19	36	120	212
Hydramnios	4	12	21	16	22	38	113
Oligohydramnios	2	3	7	11	14	20	57
IUGR	3	5	4	7	8	15	42
3. Indirect warning signs	9	20	32	34	44	73	212
4. Suspected fetal anomaly	18	21	27	30	50	79	225
Total	246	367	498	561	642	904	3218

Table 2.3. Distribution of all anomalies by referral groups, with rates of true-positive, false-negative, and false-positive diagnoses

			Total anomalies		True-positive			False-negative		False-positive	
	[n]	[n]	[n]	[n]	[n]	[n]	[%]	[n]	[n]	[n]	[n]
1. Routine examination		2569		64		59	92		5		1
Familial risk	152		11		10			1			
Exogenous insult	60										
2. Anamnestic risk		212		11		0	91		1		
Hydramnios	113		8		8						
Oligohydramnios	57		14		13			1		1	
IUGR	42		12		10			2			
3. Indirect warning signs		212		34		31	91		3		1
4. Suspected fetal anomaly		225		81		81	100				
Total		3218		190		181	95		9		2

on one hand to an increase in the number of women referred for routine screening (a general routine scanning program like that adopted in West Germany has been legislated as part of routine prenatal care in Austria) and on the other to the increasing centralization of the referral of patients with suspect findings. At present we are establishing a multistage system similar to the one described by Hansmann et al. (1985), with the levels of expertise of the examiner and sophistication of equipment increasing from Stage 1 to Stage 3.

Misdiagnoses

A total of 190 malformations (5.9%) were confirmed in the study population (Table 2.3). Of that number, 181 had been detected by ultrasound. Two cases in which a malformation had been suspected were found to be normal at delivery. In the 3035 cases declared sonographically normal, there were 9 undetected anomalies. This adds up to 11 misdiagnoses in the total population of 3218 patients (0.3%). Nine of the 190 malformations were not recognized, corresponding to a false-negative rate of 4.7%.

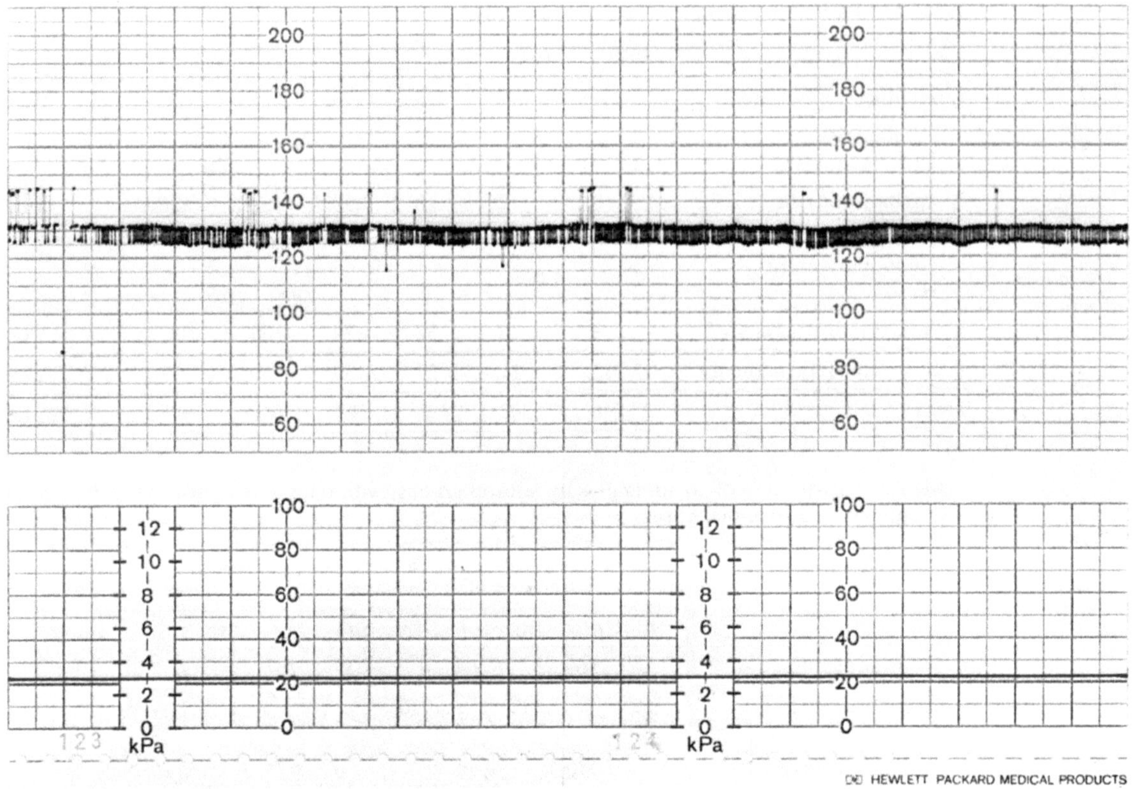

Fig. 2.9. Nonstress test in a fetus with a suspected cardiac anomaly (false-positive diagnosis, case 1, see Sect. 2.2.2, p. 14).

False-Positive Diagnoses – Analysis of Indivudual Cases

Case 1. One of the false-positive cases involved the presumptive diagnosis of a cardiac defect in a diabetic woman who was referred at 35 weeks with a nonstress test (NST) pattern not previously observed by us (Fig. 2.9). Ultrasound disclosed a grossly enlarged fetal heart with extreme dilatation of the right atrium. The foramen ovale also was markedly distended, and this finding, together with the suspicious NST, was interpreted as a cardiac anomaly. Amniocentesis was performed on the assumption that the amniotic fluid would be meconium-stained if these abnormal findings were due to fetal distress. The fluid was clear, and the patient was kept under continuous observation in the delivery suite. Serial ultrasound examinations showed no change, and the suspicion of a cardiac anomaly increased. After a detailed discussion of the case, a second amniocentesis was performed 12 h after the first. Again the specimen was clear. Two hours later the fetus expired. Autopsy disclosed diabetic fetopathy (weight 3250 g, length 52 cm), but did not show a cardiac defect or any other anomaly.

Case 2. The second false-positve case involved a patient referred at 17 weeks because of oligohydramnios that had existed for several weeks (group with indirect warning signs). Ultrasound disclosed complete absence of amniotic fluid (an-

Table 2.4. Classification of nine undiagnosed malformations by indication for referral and by organ system

	Central nervous system	Genitourinary tract	Nonimmune hydrops fetalis (NIHF)	Gastrointestinal tract	Multiple gestation	Syndromes and chromosome anomalies	Skeletal system	Tumors	Cardiac anomalies (isolated)	
1. Routine examination		2		1					2	5
Familial risk						1				
Environmental hazards										
2. Anamnestic risk						1				1
Hydramnios										
Oligohydramnios						1				
IUGR						1			1	
3. Indirect warning signs						2			1	3
4. Suspected fetal anomaly										
Total		2		1		3			3	9

hydramnios). The fetus was in a breech presentation, the BPD was 33 mm (appropriate for 14 weeks), and the OFD was 47 mm (appropriate for 17 weeks). The circumference of the dolichocephalic head was 134 mm (appropriate for 16 weeks), and the transverse trunk diameter was 33 mm (appropriate for 14 weeks). Lasix was administered, but successive examinations failed to show evidence of bladder filling; neither was it possible to visualize the renal parenchyma. The nonvisualization of the kidneys and lack of diuretic response raised the strong suspicion of bilateral renal agenesis and the pregnancy was terminated. The fetus was examined by a pathologist who was briefed on the case before conducting a very thorough examination. The fetus displayed no malformations, the kidneys appeared well formed, and microscopic examination showed no abnormalities of renal histology.

False-Negative Diagnoses – Analysis of Individual Cases

The fetal malformations were classified primarily by organ system; cases with several malformations were classified according to the major anomaly. However, since analysis by indication for referral is also important for interpreting the malformations that were missed, this relationship is shown in Table 2.4.
Looking at the anomalies by organ groups, we see that there were two undetected malformations involving the urogenital tract, one gastrointestinal anomaly, three anomalies associated with a syndrome or chromosome aberration, and three cardiac anomalies not associated with nonimmune hydrops fetalis (NIHF). Looking at the nine undiagnosed malformations by indication for referral, we find that five occurred in the "routine examination" group without anamnestic risk, and one in a patient who had a positive family history and was referred on that basis for exclusion of fetal anomalies. Three of the undiagnosed malformations occurred in the

group with indirect warning signs, one involving oligohydramnios and two involvings IUGR coexisting with oligohydramnios. No developmental anomalies were missed in the patients referred with suspected fetal anomalies.

Genitourinary Tract

Case 1. This patient was referred at 30 weeks with a breech presentation, in preterm labor. The ultrasound examination was performed as an emergency and showed no abnormalities other than anhydramnios and IUGR. A cystic mass in the lower abdomen of the fetus was interpreted as the bladder, so it was thought that a renal anomaly could be excluded. Time constraints precluded detailed anatomic screening. Tocolysis was unsuccessful, and the breech fetus was delivered by Cesarean section. The infant showed a sireniform malformation and died immediately after the delivery. Autopsy disclosed anal atresia, agenesis of the external genitalia, agenesis of the bladder and ureters, and renal cysts. These were the cysts which had been misinterpreted antenatally as the fetal bladder.

Case 2. The patient was examined at 19 weeks. Biometric parameters were appropriate for gestation, no abnormalities were described, and fetal sonoanatomy was considered normal. The child died after a spontaneous vaginal delivery; autopsy disclosed Potter type I polycystic kidneys.
Both genitourinary anomalies occurred in patients who were not thought to be at risk and underwent routine examinations.

Gastrointestinal Tract

One malformation occurred in this group. At 19 weeks the patient underwent a routine ultrasound examination which did not show any abnormalities. After an uncomplicated vaginal delivery at term, the infant required intensive resuscitation because of major respiratory problems. Examination disclosed a large left diaphragmatic hernia, which was surgically repaired (Pediatric Surgery, Dr. Henkl). The child died postoperatively from complications of pulmonary dysplasia.

Syndromes and Chromosome Abnormalities

There were three anomalies in this group that were missed by ultrasound.

Case 1. The patient had previously delivered two children with Smith-Lemli-Opitz syndrome and was referred at 17 weeks for exclusion of fetal anomalies.
We consulted the literature for the characteristic morphologic features of this syndrome and attempted to demonstrate or exclude them by ultrasound (microcephaly, nasal deformity, micrognathia, hypospadias, syndactyly of the second and third toes). Despite repeated examinations throughout the pregnancy, none of these stigmata were identified. The only remarkable finding was moderate bilateral dilatation of the renal pelves. A cytogenetic study showed normal chromosomes but a marginal elevation of the amniotic fluid alpha-fetoprotein (AFAFP). Each new examination raised our confidence concerning the exclusion of a serious defect. The patient was delivered at our hospital and immediately expressed concern that, because of its morphology, the baby was affected with Smith-Lemli-Opitz syndrome. That suspicion was subsequently confirmed.

Case 2. The patient was referred at 33 weeks because of IUGR and oligohydramnios. Both warning signs were confirmed at the first sonographic examination, at which time a cardiac abnormality was also described. However, after repeat scans, suspicion of the latter was withdrawn. Genetic amniocentesis was performed to exclude a chromosome anomaly. Before the result could be obtained, a Cesarean section was performed due to progressive deterioration of the NST. The child died postpartum from a cardiac anomaly. Subsequent chromosome analysis revealed trisomy 21.

Case 3. The patient was first examined at 34 weeks because of suspected oligohydramnios, which was confirmed by ultrasound. On evaluation of the genitourinary tract, both kidneys were visualized but bladder-filling could not be observed. No other structural abnormalities were described. Marked IUGR was also noted. Because of these findings, Cesarean section was advised. The infant died shortly after birth and on autopsy was thought to display a cryptogenic malformation syndrome, with bilateral talipes calcaneus, underdevelopment of the desmocranium (the squamous bones of the cranial vault were poorly formed and largely membranous), and hypoplasia of the urinary bladder. The urinary tract was patent and undistended, however. Both kidneys were present, of normal size, and well formed.

Cardiac Defects

Three cardiac defects were missed – two from the "routine" group and one from the group with IUGR. Both infants from the "routine" group were described as anatomically normal. One of these infants was born prematurely at 27 weeks, died soon after delivery, and was found at autopsy to have a large atrial septal defect. The other infant was delivered spontaneously at 36 weeks and also died shortly afterward. Autopsy disclosed a 20 × 15-mm defect in the upper ventricular septum and lower atrial septum with an atrioventricular canal.

In the third case, the patient was first examined at 17 weeks with suspicion of IUGR. Ultrasound showed mild IUGR and no other abnormalities. Serial scans showed a progression of the IUGR, and the patient was hospitalized at 27 weeks. A prenatal cytogenetic study was not performed. No anatomic defects were reported in any of the subsequent scans. Because of increasing growth retardation and abnormal NST, it was decided to perform a Cesarean section. The infant, which exhibited a cleft lip, palate and jaw at birth, died soon afterwards (Neonatal Unit, Prof. Dr. E. G. Huber), and was found at autopsy to have an extensive atrial septal defect and right ventricular hypertrophy. Postpartum chromosome cultures failed.

Detected Anomalies; Differentiation According to Organ System

Table 2.5 shows the 181 anomalies that were detected by ultrasound. The specific disorders are listed under each organ system and reflect the pathologic diagnoses. Although all prenatal diagnoses correctly identified the organ system involved, in 18 of the 181 cases (10%) the sonographic diagnosis was less accurate than the diagnosis from the autopsy. These cases are marked with an ■ in Table 2.5.

Prevalance of the Anomalies and Accuracy Rates

In Table 2.6 the prevalence of all 190 anomalies are indicated for the various organ systems in order of frequency, and the corresponding accuracy rates shown.

Anomalies of the central nervous system (26%) and urogenital tract (25%) show the highest frequency, followed by NIHF (13%) and gastrointestinal anomalies (12%). Much less frequent, at 8% each, are anomalies associated with multiple

Table 2.5. Distribution of the 181 detected anomalies by organ system based on the pathologic diagnosis. (■ Ultrasound diagnosis not specific enough compared with pathologic findings)

CENTRAL NERVOUS SYSTEM:		
Anencephaly	23	
Spina bifida:	8	
Hydrocephalus	14	
Dandy-Walker syndrome	2	
Intracranial cyst	1	■
Intracranial teratoma	1	■
Microcephaly	1	
Total	50	
GENITOURINARY TRACT:		
Bilateral agenesis	9	
Polycystic dysplasia (Potter type I)	3	
Multicystic dysplasia (Potter type IIA bilateral)	4	
Multicystic dysplasia unilateral + other anomalies in the opposite kidney	3	
Multicystic dysplasia isolated, unilateral	6	
Potter type III cystic kidney	1	■
Infravesical obstruction	4	
Subpelvic stenosis	6	
Obstructive megaureters	1	
Congenital megaureters	2	■
Prune belly syndrome	3	
Unilateral renal agenesis	1	■
Renal cyst	1	
Urachal cyst	1	■
Bilateral hydronephrosis (trisomy 21)	1	
Total	46	
GASTROINTESTINAL TRACT:		
Omphalocele	9	
Gastroschisis	5	
Eventration	1	
Duodenal atresia	2	
Colonic atresia	1	■
Diaphragmatic hernia	1	
Intra-abdominal cysts	3	
Total	22	
ANOMALIES IN MULTIPLE GESTATION:		
Fetofetal transfusion	7	
Acardius	1	
Thoracopagus	2	
Renal agenesis	1	■
NIHF without transfusion	2	
Multiple malformations	1	
Amelia of one foot	1	
Total	15	
SKELETAL SYSTEM:		
Osteogenesis imperfecta (Vrolik type)	1	■
Achondrogenesis	1	■
Spondylocostal dysostosis	1	■
Asphyxiating thoracic dysplasia	1	
Arthrogryposis multiplex	1	
Amelia of one foot	1	
Total	6	
NONIMMUNE HYDROPS FETALIS (NIHF):	24	
TUMORS:		
Sacrococcygeal teratoma	3	
Hamartoma of the lung	1	■
Cystic adenoid malformation of the lung	1	■
Total	5	
ISOLATED CARDIAC ANOMALY:		
Aortic stenosis	1	■
Total	1	
CHROMOSOME ANOMALIES, SYNDROMES:		
Trisomy 18	2	■
Trisomy 13	1	■
Malformation syndromes	9	
Total	12	

gestation and with syndromes or chromosome aberrations. Skeletal anomalies, fetal tumors, and isolated cardiac defects, at 3% each, account for only a small percentage of the total.

The incidence of misdiagnoses is variable. The highest rate, 75%, occurred in the group with isolated cardiac defects, followed by 20% misdiagnoses in the group with defects relating to syndromes and chromosome abnormalities.

Table 2.6. Distribution of all anomalies by organ system and corresponding rates of detection and nondetection

	Total anomalies		Detected		Not detected	
	[*n*]	[%]	[*n*]	[%]	[*n*]	[%]
Central nervous system	50	26	50	100		
Genitourinary tract	48	25	46	96	2	4
Nonimmune hydrops fetalis (NIHF)	24	13	24	100		
Gastrointestinal tract	23	12	22	96	1	4
Anomalies in multiple gestations	15	8	15	100		
Syndromes and chromosome aberrations	15	8	12	80	3	20
Skeletal system	6	3	6	100		
Tumors	5	3	5	100		
Cardiac anomalies (isolated)	4	2	1	25	3	75
Total	190	100	181	95	9	5

Table 2.7. Prevalence of all malformations by referral groups

	Examinations		Total malformations			
	[*n*]	[*n*]	[*n*]	[*n*]	[%]	[%]
1. Routine examination		2569		64		2.5
Familial risk	152		11		7.2	
Exogenous insult	60					
2. Anamnestic risk		212		11		5.2
Hydramnios	113		8		7.1	
Oligohydramnios	57		14		24.6	
IUGR	42		12		28.6	
3. Indirect warning signs		212		34		16.0
4. Suspected fetal anomaly		225		81		36.0
Total		3218		190		5.9

When we break down the total malformations (181 detected, 9 not detected) by risk groups (Table 2.7), we find a 2.5% incidence of anomalies in patients who underwent routine scanning (64 of 2569). The incidence of anomalies in the anamnestic risk group is 5.2% (11 of 212). It is interesting to note that all diagnosed anomalies were in the subgroup with a positive family history (11 of 152), and that no anomalies were detected in patients exposed to environmental hazards.

The incidence of fetal anomalies rises sharply in the group with indirect warning signs. The total incidence is 16%, with most cases occurring in the two subgroups with oligohydramnios and IUGR. The anomaly rate in patients with hydramnios was only 7.1%. As expected, the highest incidence, at 36.0% (81 of 225), was found in patients referred with suspicion of a fetal anomaly.

False-Positive Diagnoses at Referral

The 225 women who were referred with a suspected fetal anomaly were analyzed separately. It is noteworthy that in 144 of these cases, detailed screening did not confirm the original diagnosis (64% false-positive rate). Individual analysis by

Table 2.8. Patients referred between 1980 and 1985 with a suspected fetal anomaly, and corresponding false-positive rates

	1980	1981	1982	1983	1984	1985	Total
Patients referred with suspected anomaly [*n*]	18	21	27	30	50	79	225
Anomalies confirmed [*n*]	1	4	2	18	21	35	81
Anomalies excluded (false-positive referral) [*n*]	17	17	25	12	29	44	144
False-positive rate [%]	94	81	93	40	58	56	64

organ region of these 144 cases showed the following: In 85 cases (59%) the suspected anomaly was located in the region of the central nervous system. The examiner expressed suspicion of a genitourinary anomaly in 26 of the cases (18%), a gastrointestinal anomaly in 12 cases (8%), and a skeletal anomaly in 7 cases (5%). In 14 cases (10%) the diagnosis at referral did not contain any useful anatomic information on the nature of the presumed malformation. To check for a quality trend, we determined the false-positive rates for this group on a yearly basis from 1980 to 1985 (Table 2.8). We noted a decline in the false- positive rate from 94% in 1980 to 56% in 1985.

2.2.3 Discussion

Only Hansmann and Gembruch (1984) have published similar analytical data. In their study, 878 high-risk obstetric patients were grouped by referral indication for detailed screening. These authors did not have a control group without risk factors. The composition of the population in our study (Table 2.1) was only related to the referral indications and was free of selection bias. As our trend analysis demonstrates (Table 2.2), the overall frequency of ultrasound examinations in expectant mothers is rising, and an increasing number of warning signs are being observed that are prompting referral to a center for a detailed evaluation. This process is similar to the multistage concept described by Hansmann (1981). As the rising number of patients with a familial risk indicates, sonography is assuming increasing importance for the exclusion of fetal anomalies in these patients.

In patients undergoing routine examinations, the responsibility for Stage 1 screening is still frequently carried by hospital departments instead of in physicians' offices. In Austria this population is diverse, in that many patients have had previous scans outside the hospital, most of which have not been documented. There are other patients for whom the ultrasound examination in the hospital department represents their initial examination. There is a wide range of gestational age at the time of the first examination and dates often cannot be accurately established due to late referrals. We therefore decided not to group our patients by gestation or set cutoff points like the 23-week limit used by Hansmann and Gembruch (1984). It is anticipated that this population will move increasingly from the hospital to the office setting as the mandatory screening program is instituted.

The retrospective collection of data in the patients declared normal cannot guarantee the exclusion of minimal intracorporeal defects in their offspring. However, it is reasonable to assume that the frequent pediatric examinations and clinical follow-up were sufficient to exclude major anomalies in these newborns.

When anomalies are detected by ultrasound, the presumptive diagnosis has been shown to influence significantly the thoroughness of the examination of surviving infants and postmortem examinations in induced abortions and perinatal deaths (Födisch 1982; Rehder 1982; Födisch and Knöpfle 1984).
The total of 11 misdiagnoses may seem relatively small but cannot be compared to the success rates of authors who have been examining patients for several years at a Stage 3 facility (Hansmann and Gembruch 1984; Hansmann et al. 1985). The more highly differentiated the diagnostic "challenge" from the referring physician, the higher the likelihood of missing the diagnosis. This particularly applies to the diagnosis of neural tube defects before 19 weeks. We were faced with this question in only 1 of the 8 diagnosed cases of spina bifida (Table 2.5) prior to 19 weeks. Even in this case, the only indication for detailed examination was an elevated AFAFP observed in connection with a cytogenetic study performed because of advanced maternal age.
Both false-negative cases resulted from failure to implement further diagnostic measures, due to lack either of diagnostic instrumentation (Doppler ultrasound for cardiac defects) or of experience with invasive diagnostic procedures (amniotic fluid replacement in oligohydramnios). If we analyze the nine undetected malformations, we again find that inadequate diagnostic follow- through was to blame in at least some of the cases. This particularly applies to the false- positive [sic] diagnoses in the high-risk population. The exclusion of Smith-Lemli-Opitz syndrome (case 1) should not have been assumed, as the examiner had no previous visual experience with the sonomorphologic features of this condition. In the two other undiagnosed cases from the group with syndrome- or chromosome-related anomalies, evaluation was made difficult by the reduced amniotic fluid volume. Also, case 2 was not evaluated further for cardiac defects despite the original impression of a cardiac anomaly. In this case the late referral at 33 weeks limited the time available for an adequate diagnostic workup. The same is true of case 3 from this group.
The failure to detect genitourinary anomalies in case 1 (obstructive uropathy, sireniform malformation) was due to an incomplete examination procedure (Staudach 1982). In case 2 (Potter type I cystic kidneys), it must be considered that the amniotic fluid volume may still be normal before 23 weeks (Weiss et al. 1981), and that very small cysts may not be visualized with older equipment. In three cases with Potter type I polycystic dysplasia (Table 2.5), a correct diagnosis was subsequently established. The major finding in all cases was echogenic, solid-appearing kidneys, and confirmation was provided by determining the ratio of the mean kidney circumference to the abdominal circumference (Grannum et al. 1980). Diagnosis was facilitated by the use of 5-MHz transducers, whose superior resolution allowed for delineation of minute cystic structures.
In the cases of the undetected diaphragmatic hernia, serial scanning and organ identification were both incomplete, since at the time of surgery there was marked carciac displacement toward the right side by intrathoracic bowel loops.
Of the three undiagnosed cardiac defects, the morphologic changes should have been detectable in case 2 (atrioventricular canal). In case 3 the IUGR and oligohydramnios were not considered alarming enough to justify cytogenetic study. A definitive diagnosis was never established in this case, for even the postpartum cytogenetic evaluation was unsuccessful.
The 181 correctly diagnosed anomalies include 18 cases in which the sonographic diagnosis did not provide the degree of detail achieved by the clinical or postmortem examination. However, this lack of specificity did not affect perinatal

management. The major problem in this group was the difficulty of interpreting anomalies that were being seen for the first time. In such cases the examiner is unable to relate the sonographic structural abnormalities to a visual image of the corresponding pathology in his memory. This fact is a strong argument in favor of the multistage concept of Hansmann (1981), which recognizes that rare anomalies can be diagnosed consistently only by referring these cases to a tertiary facility. Thus, in three fetuses with skeletal anomalies (Table 2.5), it was possible only to identify the affected region and diagnose the presence of an osseous defect; a specific prenatal diagnosis could not be established. In the three cases of chromosome disorders in our series, sonomorphologic features raised the suspicion of an anomaly, but cytogenetic studies were needed to corroborate the finding. The importance of adjunctive cytogenetic findings in obstetric patients with sonographic abnormalities cannot be overemphasized (Staudach et al. 1984). Cytogenetic evaluation was omitted in a case of bilateral hydronephrosis and in a confirmed case of duodenal atresia, and both infants were found at birth to have trisomy 21.

The order of frequency of the diagnosed malformations by organ system is consistent with the data reported by other authors (Winter 1981; Bernaschek et al. 1980; Hansmann and Gembruch 1984; Hobbins et al. 1979; Hansmann et al. 1985). The relatively rapid rise in the percentage of diagnosed urinary tract anomalies is noteworthy. If this trend continuous, this group may eventually rank first in the spectrum of fetal anomalies that are diagnosable by ultrasound.

Looking at the prevalence of anomalies by referral group, we find that only 7.2% occurred in patients with a familial risk. This is consistent with the results of Hansmann and Gembruch (1984), who found only a 4.1% recurrence rate in their obstetric patients with positive family histories. We find similar numbers in the group exposed to environmental hazards. We did not observe any malformations in this group, while Hansmann found only 1 anomaly (a neural tube defect) in a comparable population of 205 patients (= 0.5%). In the risk group with indirect warning signs, anomalies are most prevalent (about 25% incidence each) in the subgroups referred for oligohydramnios and IUGR. It should be noted that these patients were grouped by referral indication, and that not all the patients referred for "oligohydramnios" were actually found to have that condition. In the group referred with a suspect fetal anomaly, the incidence of false-positive diagnoses decreased markedly during the interval of the study (Table 2.8). It is expected that this trend will continue with the increasing number of patients screened at Stage 1 level.

It appears significant that 33% of all the anomalies (64 of 190) occurred in the group undergoing routine scanning. Only 5.8% of all anomalies (11 of 190) occurred in patients with positive histories. Finally, it is noted that 94.2% of anomalies were detected as a result of voluntary screening, underscoring the importance of routine prenatal ultrasound examinations (mandatory screening) as a tool for the diagnosis of fetal anomalies.

3 Examination Procedure

3.1 Introduction

To develop expertise in prenatal sonography and especially in the evaluation of fetal anatomy, the examiner must be able to:

1. relate three-dimensional transducer movements on the maternal abdomen to the two-dimensional sonographic image and mentally construct a three-dimensional structure from the sum of the individual, visually recorded images;
2. recognize anatomic structures on the basis of sonographic morphology and varying shades of gray;
3. rapidly assemble the sum of analytically recorded anatomic data into a comprehensive, anatomically correct image.

3.2 Orientation and Examination Setup

When dealing with problems of sectional anatomy, it is necessary that the terms are defined. The following cardinal body axes form the basis for anatomic orientation: the craniocaudal (longitudinal) axis, the anteroposterior (AP) axis, and the right-left axis. Because two lines are needed to define a plane, every sectional plane lying in one of the three cardinal directions in space must be defined by two of the axes named above. Owing to the bilateral symmetry of the human body, the only plane that divides the body into symmetrical halves is the midsagittal plane, which is defined by the craniocaudal axis and the AP axis (Fig. 3.1). The plane defined by the AP axis and right-left axis is called the transverse plane (or horizontal plane). The third major plane is the frontal (coronal) plane, defined by the craniocaudal axis and the right-left axis. The "tilting" of sections can occur on either of two axes.
As explained in the section on Trunk Biometry (Chap. 8), measurements of the trunk can be distorted by tilting of the transverse plane. Tilting the plane on the AP axis increases the apparent value of the transverse trunk diameter (Fig. 3.2 a), while tilting on the right-left axis increases the apparent AP diameter (Fig. 3.2 b). Anatomic orientation is based on these same axes, using the term "craniocaudal" in reference to the longitudinal axis, "anteroposterior" in reference to the AP axis, and "right-left" when referring to the right-left axis. The terms "lateral" and "medial" should be used only in reference to directional relationships within a body half.

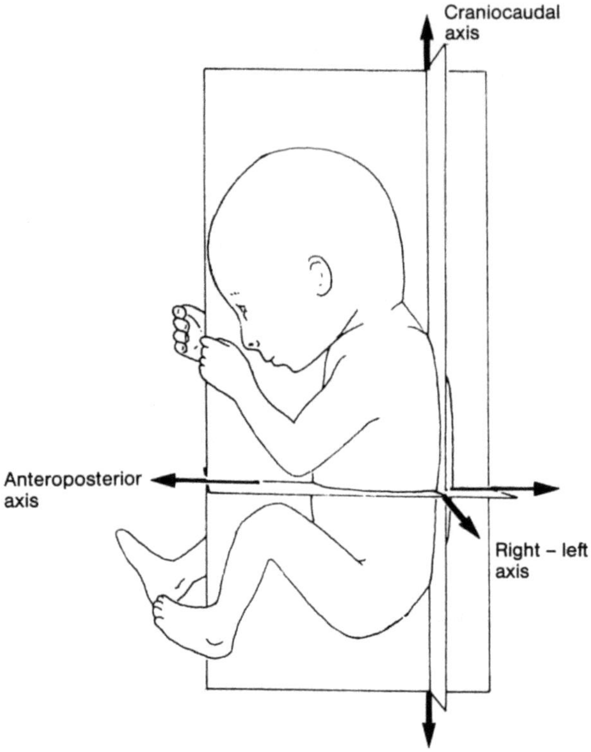

Fig. 3.1. Schematic diagram of the cardinal axes and cardinal planes of the body

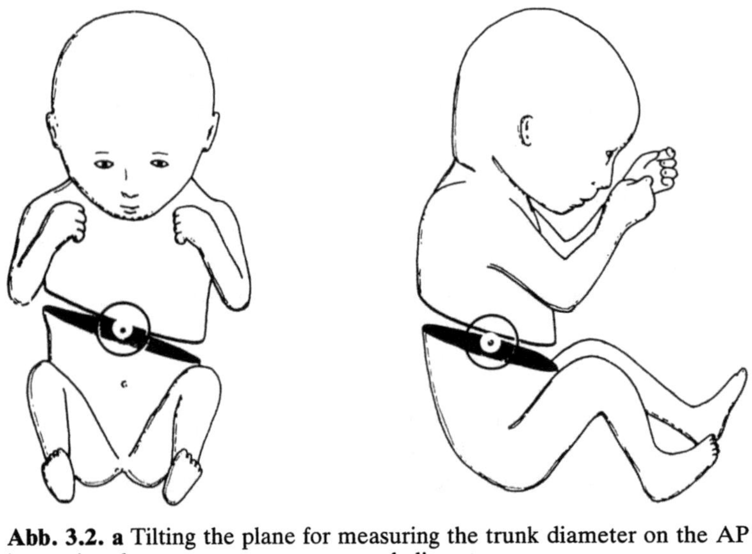

Abb. 3.2. a Tilting the plane for measuring the trunk diameter on the AP axis has the effect of increasing the apparent transverse trunk diameter

Fig. 3.2. b Tilting the plane for measuring the trunk diameter on the right-left axis has the effect of increasing the apparent AP trunk diameter

Besides problems of nomenclature, the identification of scanning planes in obstetric sonography is made difficult by variations in the setup used for the examination. For postpartum ultrasound, the examiner generally sits facing the patient. Many examiners use this same position for antenatal scans. For several years we have been using a reverse of this setup, with the examiner facing the same direction as the patient while looking at the monitor screen. The rationale for this setup is largely psychological. The parallel gaze of the patient and examiner reinforces their common interest in the sonographic image and also makes it easier for the examiner to point out specific features. The common "face-to-face" arrangement does little to reinforce the sense of joint scrutiny of the image, even if the monitor is tuned to the woman. In a study by Reading and Cox (1982), women who were shown the fetus on the monitor screen and given explanations about what they were seeing responded much more positively than did women who were not given this information.

As it may be difficult for the observer to appreciate anatomic relationships on some sonograms without knowing the viewing angle of the examiner at the time the scan was made, we have indicated the viewing direction in a number of cases by means of diagrams or text explanations.

A word about patient positioning should be said in connection with the examination setup described above. Best results are obtained by having the patient recline on a low scanning bed with an adjustable back rest, while the examiner sits on a swivel chair which makes it easy to alternate between viewing the screen and talking with the patient.

In this book we follow the usual rules for directional relationship on sonograms: On transverse scans, structures in the left part of the maternal abdomen appear on the right side of the image; on longitudinal scans, the right side is caudal. It should be added, however, that the physician practising obstetric ultrasound must be well versed in the different views associated with breech and cephalic presentations when identifying sectional planes in fetuses (e.g., viewing a cranial section from the caudal aspect or a caudal section from the cranial aspect).

3.3 General Survey

The examination should always begin with a general survey (we often start the examination with linear transducers, especially in the third trimester) that includes the uterus (position, size, shape, tumors). If the maternal bladder is full, we include visualization of the cervix as a routine part of the survey (Klug et al. 1985). Working with the aforementioned author, we were able to demonstrate the cervix without difficulty in 343 patients between 5 and 32 weeks (Figs. 3.3 and 3.4). This step is particularly important in cases where cervical incompetence has been diagnosed from a vaginal examination alone (Fig. 3.4). Another indication for ultrasound cervicometry exists in women who would suffer extreme emotional upset from a vaginal examination. Following the assessment of the uterus, or in connection with it, we recommend localization of the placenta. Details pertaining to placental visualization and morphology are outside the scope of this text.

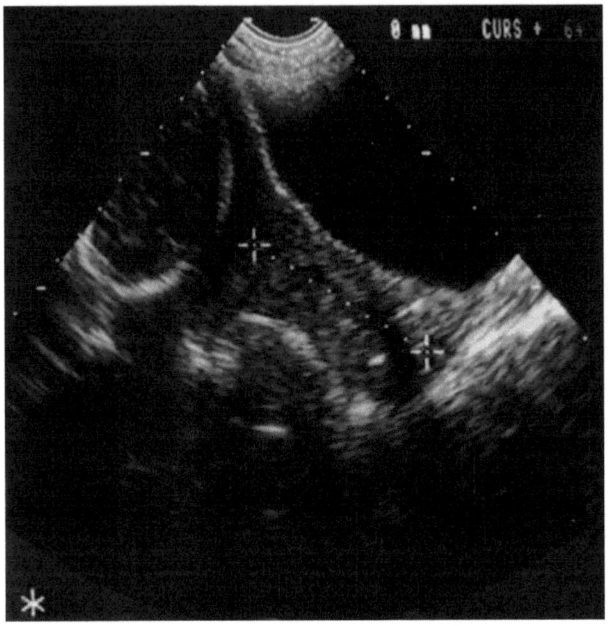

Fig. 3.3. Measurement of the cervix in pregnancy with a full bladder. The cervix is well delineated. The cervical length (64 mm) is measured from the internal to the external os

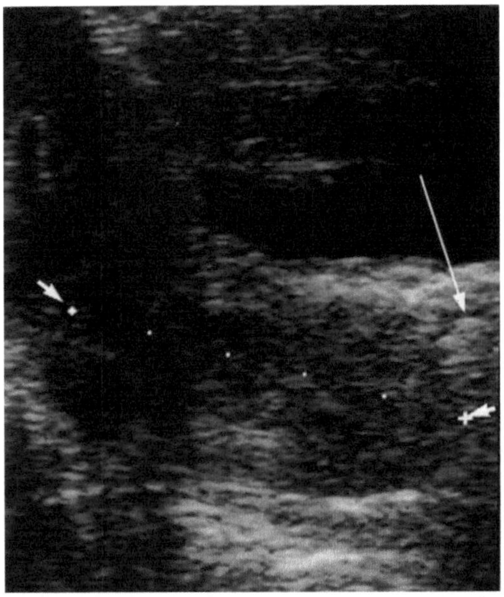

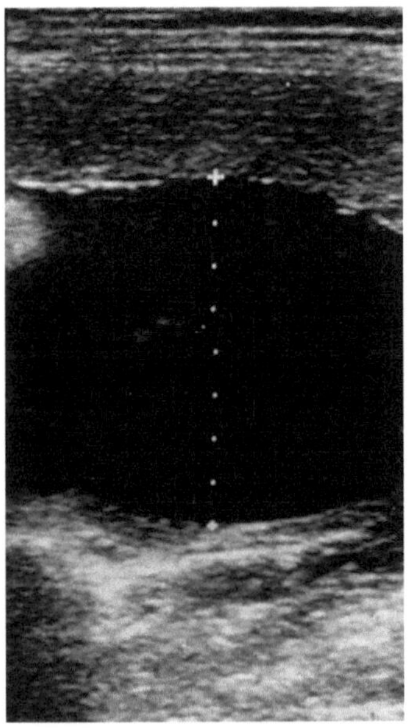

Fig. 3.4. (left) Measurement of the cervix in pregnancy. The long arrow marks the examiner's finger in the anterior fornix. Vaginal examination indicated a short cervix, but the sonogram demonstrates a well-formed cervix 50 mm in length

Fig. 3.5. (right) Semiquantitative measurement of the amniotic fluid volume. The anteroposterior diameter of the largest amniotic fluid pocket is measured

3.4 Evaluation of Amniotic Fluid Volume

By this time the examiner will already have formed an impression of the amniotic fluid volume. Thus, if the posterior uterine wall or placenta appear particularly clear and echogenic, the experienced examiner will automatically associate this with an increased amniotic fluid volume, and conversely, poor visualization of these structures will suggest a paucity of fluid. An exact, quantitative determination of amniotic fluid volume is not yet possible in routine examinations, although Chamberlain et al. (1984 a, b) used semiquantitative parameters to show that abnormalities of amniotic fluid volume bore a definite relationship to perinatal outcome. In their technique, serial scans of the uterine cavity are performed to find the largest pocket of amniotic fluid, whose maximum vertical and transverse diameters are determined. The depth of the amniotic fluid pocket is then measured at a right angle to the uterine wall (Fig. 3.5) and placed in one of four qualitative categories. If the maximum AP diameter is less than 1 cm in the largest fluid pocket, the amniotic fluid volume is considered decreased. The range of 1–2 cm is classified as marginal, 2–8 cm as normal, and a value in excess of 8 cm signifies hydramnios. In all cases where the AP diameter was greater than 1 cm, the concurrent measurement of the transverse diameter also exceeded 1 cm. The authors (Chamberlain et al. 1984 a) evaluated the results from more than 7000 high-risk pregnancies. In cases where the maximum amniotic fluid volume in the designated planes had a diameter of less than 1 cm, perinatal mortality showed a highly significant increase compared to patients with normal values.

It should be added that these results pertain to pregnancies beyond 24 weeks' gestation. Since 1984 we have incorporated this parameter into our diagnostic workup of selected patients, but so far we have been unable to establish its predictive value. Fig. 3.5 illustrates the measurement of the AP diameter of the largest amniotic fluid pocket observed at 29 weeks. The 8.1-cm value is just above the normal range defined by Chamberlain et al. (1984 a. b).

Table 3.1 presents a list of conditions that are associated with an abnormal amniotic fluid volume (Staudach 1984). This list is by no means exhaustive, but it should be

Table 3.1. Possible causes of abnormal amniotic fluid volume

Polyhydramnios	Oligo- or anhydramnios
Anencephaly	Renal agenesis
Meningomyelocele	Microcystic renal dysplasia
Hydrocephalus	Multicystic renal dysplasia
Esophageal atresia	Obstructive uropathy
Duodenal atresia	Abnormal airway development
Cardiac decompensation	Abnormal lung development
Fetofetal transfusion	Syndromes
Nonimmune hydrops fetalis	Chromosomal anomalies
Atresia of the airways	Placental insufficiency
Partial urinary tract obstruction	Lethal loops of cord around fetus
Chromosomal anomalies	Rupture of the membranes
Syndromes	
Intrauterine infections	
Rh incompatibility	
Diabetes	
Idiopathic	

of help in correlating abnormalities of fluid volume with possible pathologic states. Information on amniotic fluid volume in normal pregnancies may be found in Queenan and Thompson (1972). Manning et al. (1981), Hill et al. (1983), Philipson et al. (1983), and Crowley et al. (1984) have emphasized the importance and predictive value of amniotic fluid volume determinations. Halperin et al. (1985) were able to show that even without quantitative parameters, an experienced examiner can correctly estimate the amniotic fluid volume by empirical means.

3.5 Fetal Presentation and Position

Before the examiner can analyze fetal anatomy, he must know the position of the fetus in the uterus and confirm that the fetus is alive (cardiac action, movements). Once the examiner has established the fetal position, we recommend that he imagine himself assuming the position of the fetus and mentally transpose further anatomic relationships onto his own body. The fetal presentation (breech, cephalic) and position (relationship of the fetal presenting part to the right or left maternal side) will have a major bearing on the detailed examination that is performed later. Thus, for example, an occiput anterior position greatly limits the ability to evaluate the fetal heart and profile. A breech presentation generally makes it easier to demonstrate the intracranial anatomy, and sagittal and frontal scans are far more easily obtained than in the cephalic presentation. Göttlicher et al. (1981) gathered data on the prevalence of longitudinal (breech and vertex presentation), transverse, and oblique lies in a large obstetric population. They found significant differences between primiparous and multiparous women and computed the probability for the version of a breech into a cephalic presentation in relation to gestational age. This probability was 44% at 28 weeks and declined to 23% by 32 weeks. We personally determined the prevalence of breech, cephalic, and unstable presentations during the course of 1213 consecutive ultrasound examinations performed between weeks 16 and 34 (Fig. 3.6) By weeks 16 and 17, 50% of the fetuses

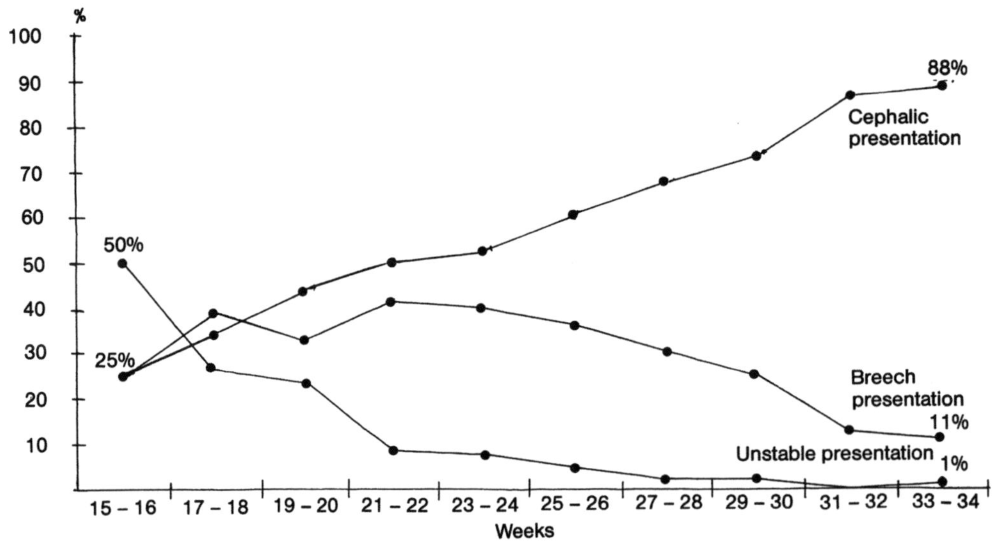

Fig. 3.6. Distribution of fetal presentation vs. gestational age between 15 and 34 weeks

examined were in an unstable position (oblique to the maternal longitudinal axis with potential for rotation, transverse lies changeable by external manipulation, spontaneously observed changes of presentation in utero). The prevalence of breech and cephalic presentations at this stage was 25% each. As gestational age increased, the prevalence of cephalic presentation rose to 88% by week 35, with an 11% frequency of breech presentation. To summarize the data obtained for the critical screening period between weeks 16 and 23, 40% of the 480 fetuses examined were cephalic, 36% were breech, and 24% unstable presentations.

3.6 Fetal Body Surface

Once the examiner has determined the fetal presentation and position, he proceeds to an overall assessment of the fetal body surface. This is done by rotating the transducer about the craniocaudal axis, as far as the maternal abdomen allows it. It should be noted that initially inaccessible scan planes often can be obtained simply by adjusting the maternal position. The fetal body outline is visualized and attention being given to structural defects or protrusions. The major anomalies that can be detected by fetal surface screening, along with their characteristic features, are listed in Table 3.2. Most of these anomalies are closure defects, which can be detected only if the fetal body is completely outlined. Difficulties can arise if parts of the fetal body are obscured by acoustic shadowing from the extremities, or if the fetus is lying directly against the uterine wall. Induced fetal position changes, the use of "natural acoustic windows", and maternal repositioning can be helpful in these instances.

Another important aspect of the evaluation of the fetal outline is the assessment of the body proportions, including the relationship between the skull and facial skele-

Table 3.2. Major developmental anomalies detectable by sonographic evaluation of the fetal body surface

Structural defects of the cranial vault, absence of the calvarium, "goggle-like" orbits	Anencephaly
Structural defects in hyperextended cranium, spinal deformity	Iniencephaly
Sac with solid and/or cystic contents extending from a defect in the cranial vault	Encephalocele Occipital meningocele
Cystic structures in the neck region	Hygroma cervicis
Structural defects of the spine with uni- or multilocular cysts	Spina bifida
Caudal tumor with solid and cystic contents	Sacral teratoma
Defect in the thoracic region with ventral displacement of the heart	Ectopia cordis
Anterior abdominal wall defect with extruded bowel; umbilical cord enters abdomen separate from mass	Gastroschisis
Anterior abdominal wall defect with herniating bowel loops and organs (liver, spleen) within membrane; mass occurs at insertion of umbilical cord	Omphalocele
Anterior lower abdominal wall defect with protrusion of a "cystic structure"	Ectopia vesicae

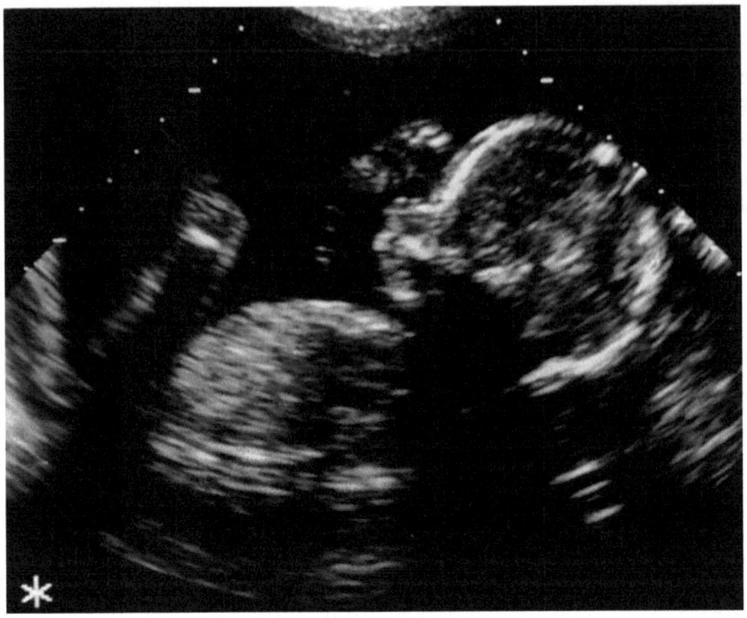

Fig. 3.7. Overview of the fetus. Midsagittal scan through the head, face, and trunk permits evaluation of the facial profile, body surface, and relationship between the various structures

ton, the relation of skull size to trunk size, and the ratio of chest and abdomen size. Often an abnormal ratio can be detected at this stage even before biometry has been performed. Figure 3.7 shows an ideal scan for evaluating the anterior body surface and body proportions.

3.7 Examination of Sectional Anatomy

Examination of the fetal sectional anatomy for the purpose of evaluating intracorporeal structures should likewise begin by locating standard planes defined by the cardinal body axes. The three major planes for linear scanning are shown schematically in Fig. 3.8 a–c. Generally at this stage we try to follow a standard approach. We begin with posteroanterior sagittal scans, paying special attention to the skull outline, neck region, and spine (see Chap. 5). The kidney is the key target organ in this plane on account of its anatomic relations. A frontal section often can be obtained by rotating the transducer 90° on the craniocaudal axis. This scan permits an assessment of fetal body symmetry and aids further in the structural analysis of the spine. The position and size of the fetal stomach also can be determined in this plane, and the symmetry and position of the kidneys can be assessed (see Chap. 9). The indirect demonstration of the diaphragm often serves to demarcate the thorax from the abdomen. The fetal hip and shoulder areas can also be identified in this plane.
From here the transducer can be rotated another 90° on the longitudinal axis to obtain a sagittal section from the anterior side. Because the back of the fetus is

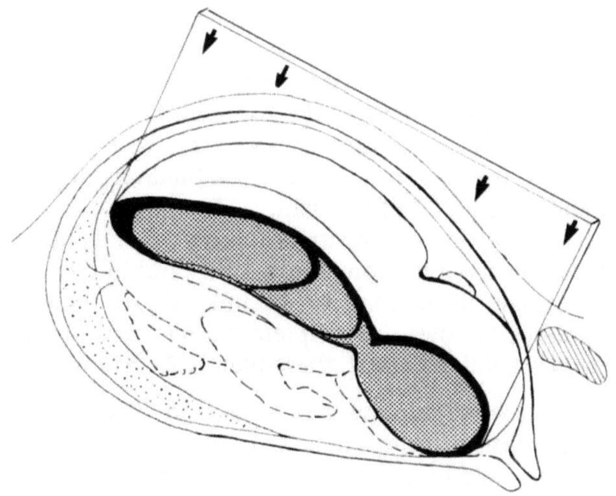

Fig. 3.8. a Posteroanterior parasaggital scan

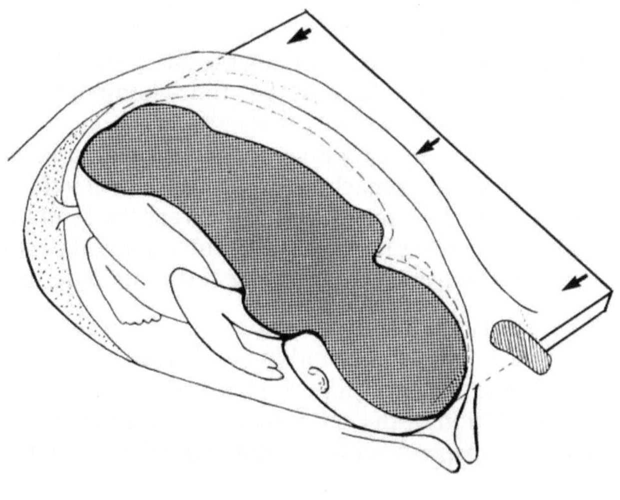

Fig. 3.8. b Frontal scan

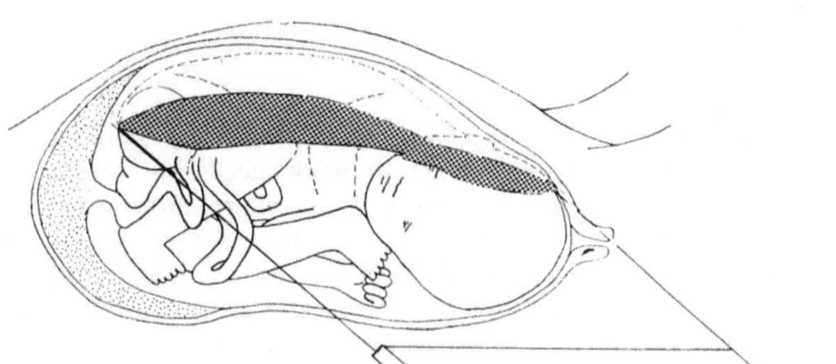

Fig. 3.8. c Anteroposterior parasagittal scan

posterior, the "view" into the thorax and abdomen is unobstructed by acoustic shadowing from the vertebrae and ribs. In this plane the examiner can inspect the facial profile, check the integrity of the anterior body surface, assess the ratio of the thorax and abdomen, and reexamine the diaphragm from a different angle. The liver, stomach, and umbilical vein can be brought into view by moving the transducer to the right or left. The fetal bladder, thighs, and genitalia can be identified in the caudal part of the scan.

After surveying the sectional anatomy in the planes indicated, we rotate the transducer 90° from the longitudinal axis and initiate a transverse series. Of the many transverse planes that can be imaged, we recommend selecting a few standard planes that allow for a complete examination at a minimum of time (Fig. 3.9 a, b). From cranial to caudal, these planes are as follows:

1. Transverse (biparietal) cranial scan. The symmetry of the section is assessed on the basis of head shape and centering of the midline echo between the cranial walls. We first obtain transverse plane 1 (see Sect. 4.3) and evaluate the morphology and size of the lateral ventricles.
2. The transducer is moved caudally to obtain the reference plane for fetal cephalometry. Following a gross assessment of the intracranial structures in this region (frontal horn, cavum septi pellucidi, thalamus, occipital horn), which serve as criteria for identifying the reference plane, appropriate measurements can be taken for determining the BPD and OFD.
3. The third target plane ist the "four-chamber view" of the heart. Its features are described in Chap. 7.
4. The fourth target plane serves to perform trunk biometry, and also permits evaluation of the stomach, liver, and umbilical vein.
5. The next lower scan will demonstrate the fetal kidneys.
6. Plane six serves to visualize the bladder and evaluate the thigh area. The femur length often can be accurately determined on this scan.

The diagnosis of fetal structural anomalies is generally based on the demonstration of parenchymal defects and organ displacements, the identification of cystic masses, and the presence of distended cavities and fluid collections.

On completion of the serial scans, even a routine ultrasound examination should include an inspection of the fetal extremities (see Chap. 10).

3.8 Errors

A major source of error is a disorganized scanning technique. The greater the distance between scans performed in quick succession, the more difficult it is for the examiner to establish the proper anatomic relationship and maintain the orientation. An erratic technique also compounds the risk that organ systems crucial to the diagnosis and overall evaluation will be missed. Generally the best results are obtained when the examiner takes time and moves smoothly and systematically from one area to the next. The experienced examiner will not wait to see which anatomic structures present in the next image plane, but will tend to anticipate the upcoming sectional anatomy so that he can readily confirm normal morphology or recognize an anomaly when the corresponding area is reached.

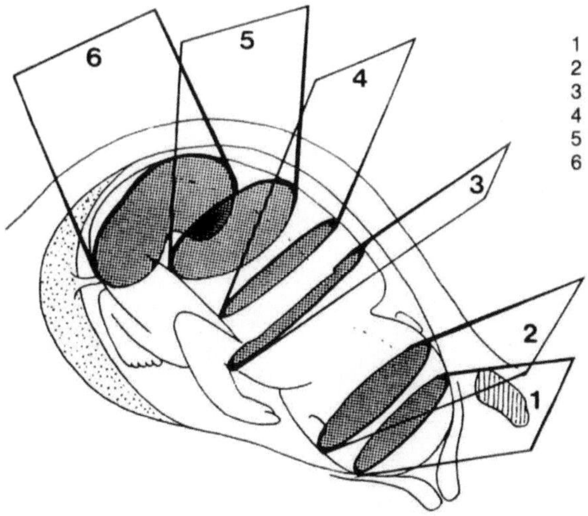

Fig. 3.9. a Schematic diagram of the six transverse scan planes used for basic evaluation

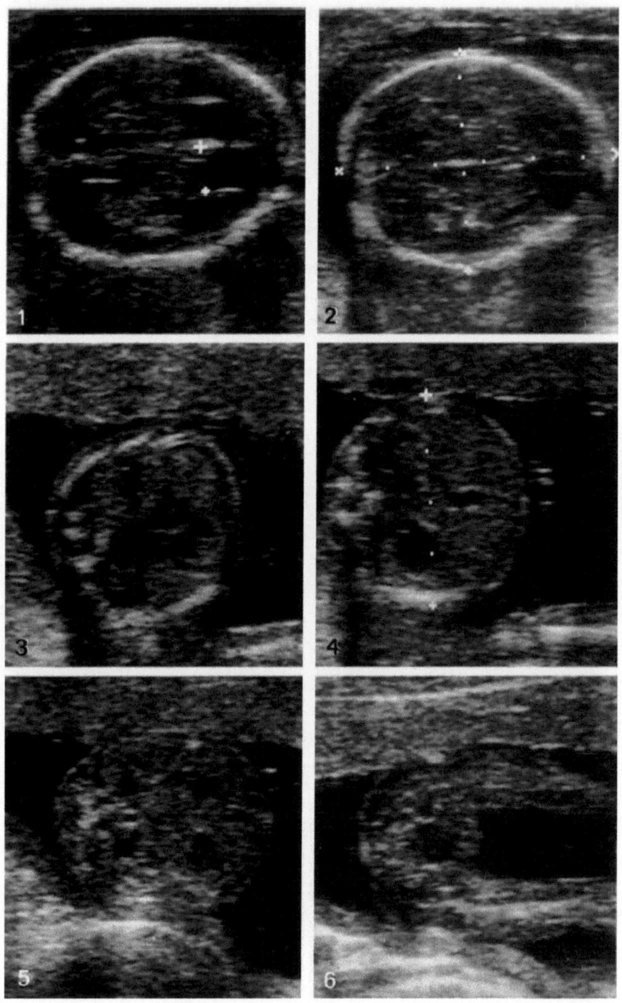

Fig. 3.9. b Illustrative sonograms of the standard transverse planes. *1* Scan through the lateral ventricles, demonstrating the ventricular width. *2* Reference plane for measurement of BPD, OFD, and head circumference. *3* Transverse scan through the thorax, showing "four-chamber view" of the heart. *4* Reference plane for trunk biometry. *5* Transverse scan demonstrating the kidneys. *6* Transverse scan demonstrating the bladder and lower extremities

4 The Head

4.1 Introduction

Sonographic visualization of the fetal head is crucial in obstetric ultrasound. On the one hand, the various quantitative parameters of the fetal skull are a cornerstone of fetal biometry; on the other, malformations of this region rank first among the fetal developmental anomalies that are detected (Winter 1981; Hobbins et al. 1983; Hansmann et al. 1985). This is illustrated by our own study population, in which the most frequent anomalies were those involving the central nervous system (see Table 2.5).

The physician who practices obstetric ultrasound must have a sound understanding of fetal neuroanatomy and its sonographic correlates in order to differentiate normal findings from pathology. Standard descriptive anatomic terms are used to designate the planes of section (Fig. 4.1). The position of the fetus within the uterus will determine in part the accessibility of the desired planes. Fetuses showing a cephalic presentation are most easily scanned in the transverse and frontal planes. Fetuses in a breech presentation generally can be scanned in the sagittal plane as well. As mentioned in Chap. 3, determination of the fetal presentation is a necessary prelude even to examinations of the fetal head and brain. The degree of flexion or extension of the head also is noted in relation to the trunk axis. This is important in locating the necessary planes.

We recommend starting the fetal head examination with an inspection of the superficial bony structures. Care is taken not to mistake physiologic "gaps" (fontanelles, sutures) for pathologic defects (anencephaly, encephalocele, myelocele). Next the cranium and facial skeleton are compared, preferably by means of a midsagittal profile view. This scan is most easily obtained by starting with a transverse scan and rotating the transducer in place 90° about the AP head axis (Fig. 4.2). This evaluation is followed by selective scans that demonstrate the intracerebral anatomy and establish the reference planes necessary for biometry. We recommend concluding the fetal head examination by delineating the facial soft-tissue structures (eyes, nose, mouth), as this will provide the examiner with necessary practice in visualizing that area and strengthen the ability to distinguish normal from pathologic findings.

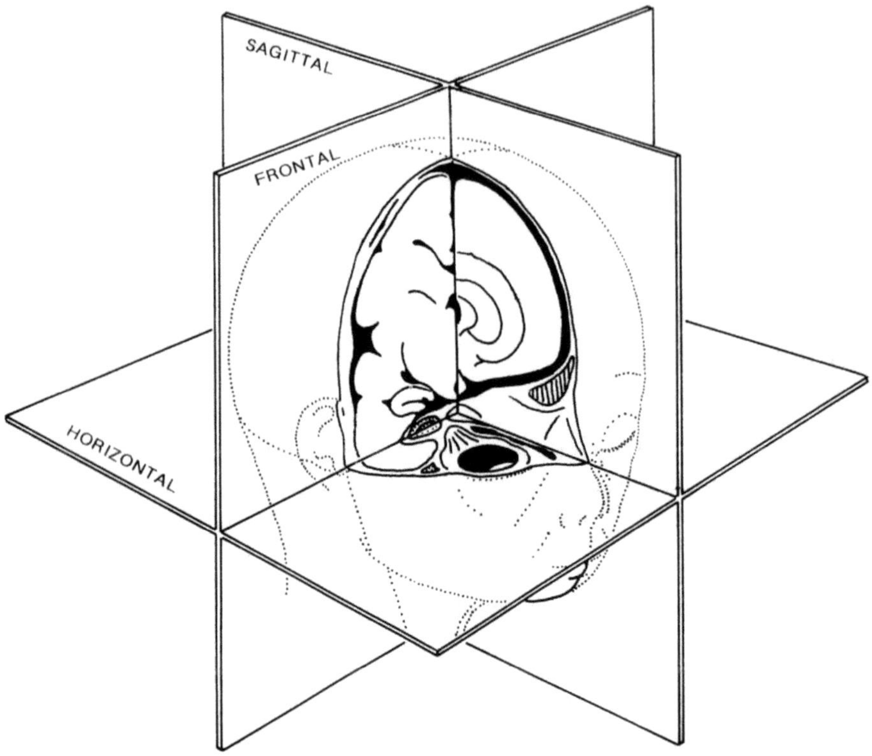

Fig. 4.1. Schematic diagram of the major planes of section of the head

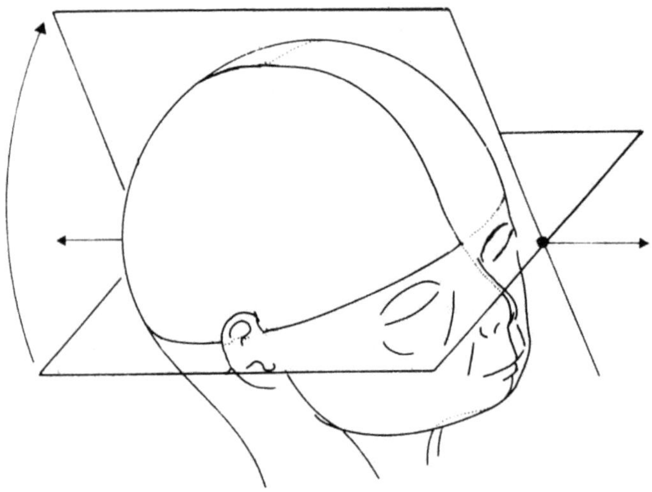

Fig. 4.2. The profile is delineated by rotating the transducer 90° about the AP axis from the transverse plane

Table 4.1. Earliest detection of ossification centers in the embryonic head (numbers correspond to key numbers in the figures)

Bone	Can be identified from:	Key number
Mandible	7 weeks	1
Maxilla	7 weeks	2
Frontal bone	9 weeks	3
Parietal bone	10 weeks	4
Occipital bone	10 weeks	5
Sphenoid bone	11 weeks	6
Temporal bone	11 weeks	7
Zygomatic bone	11 weeks	8
Nasal bone	10 weeks	9

4.2 Cranial Bones, Fontanelles, Sutures

The complex developmental pattern of the fetal skull has been traced radiologically by Davies and Davies (1962) and by Kier (1971). Ultrasound studies on the development of the cranial bones have not previously been published. Sonographic delineation of the cranial contour relies mainly on density differences relative to the ambient medium. The acoustic impedance difference between the amniotic fluid and connective tissue structures and the early delineation of the fetal outline by the appearance of ossification centers permits recognition of the head as early as 7 to 8 weeks (Fig. 4.3 and 4.4).
The first sonographically visible ossification centers appear near the end of the first trimester. The results of our own investigations are shown in Table 4.1, which lists the cranial bones relevant to obstetric ultrasound and the times at which their ossification centers can first be seen. To aid in the orientation on the sonograms and to avoid cluttering the pictures with arrows, the numbers listed in the table correspond to the key numbers used in the illustrations. From 10 weeks on, the fetal head exhibits an increasingly distinct outline in sagittal section, and the cranial bones can be individually demonstrated and identified. Figures 4.5 and 4.6 show the osseous structures that are identifiable at 11 weeks in different planes. Progressive ossification has clearly defined the cranial outline, and sutures and fontanelles can be identified for the first time (Fig. 4.6). Ossification of the frontal bone starts in the supraorbital area and is responsible for the early delineation of the orbits (Fig. 4.8). Frontal scans between 10 and 16 weeks show progressive ossification of the frontal bone with increasing demarcation of the anterior fontanelle (Figs. 4.7 and 4.8). Figures 4.9 and 4.10 illustrate the progressive ossification of the frontal, parietal, and occipital bones. These bones show only patchy ossification at 13 weeks, but by 16 weeks the cranial vault displays a continuous contour on the sagittal scan, interrupted only by the fontanelles and sutures (Fig. 4.10). An exact midsagittal scan (midline scan), obtained either by accident or by a selective search, which passes through the still relatively wide frontal suture, anterior fontanelle, sagittal suture, and posterior fontanelle will demonstrate no osseous portions of the skull except for the occipital bone (Fig. 4.11). Even at this early stage, intracranial structures can be seen very clearly on scans performed through this "acoustic window" (see Sect. 4.3).

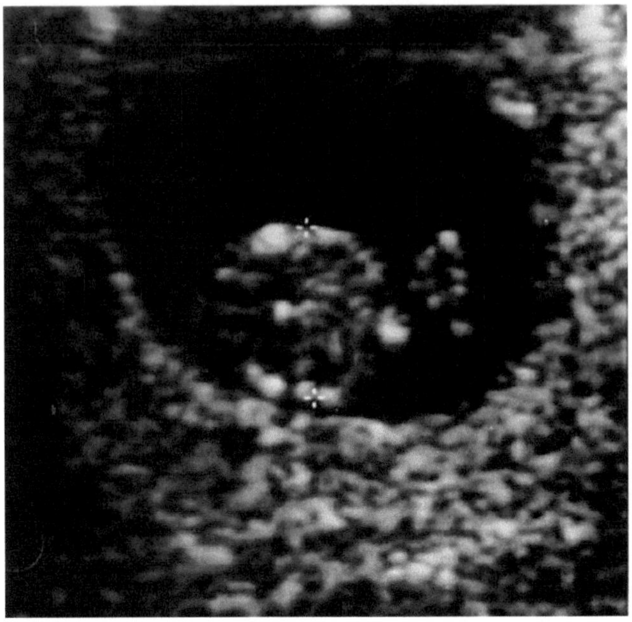

Fig. 4.3. Measurement of the BPD in an 8-week embryo

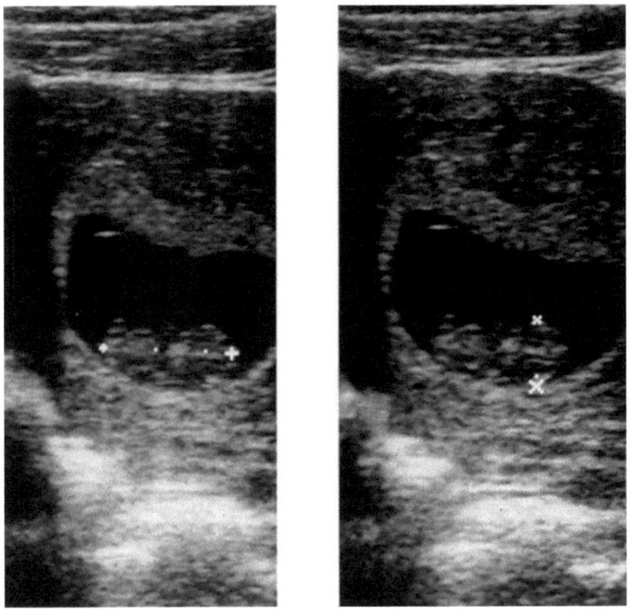

Fig. 4.4. The head and trunk of this 8-week embryo are clearly identifiable as separate structures (reference marks for CRL and BPD are shown)

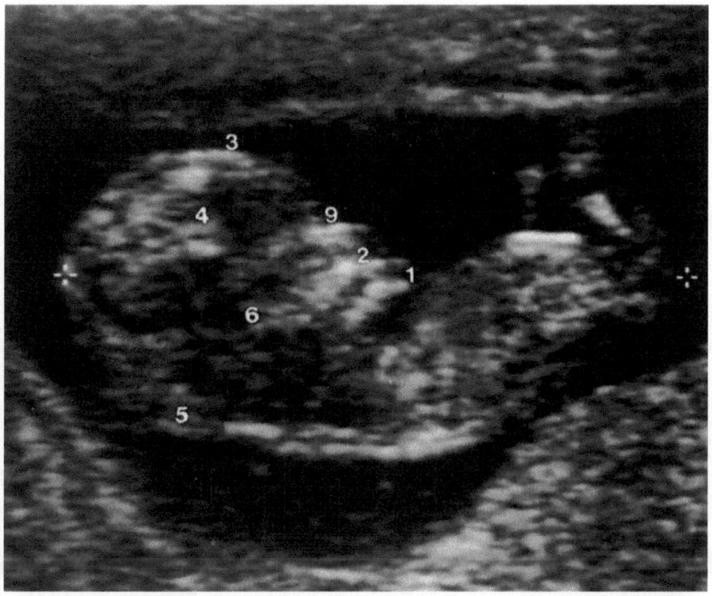

Fig. 4.5. Midsagittal scan through an 11-week fetus. Individual cranial bones can be distinguished by their ossification centers

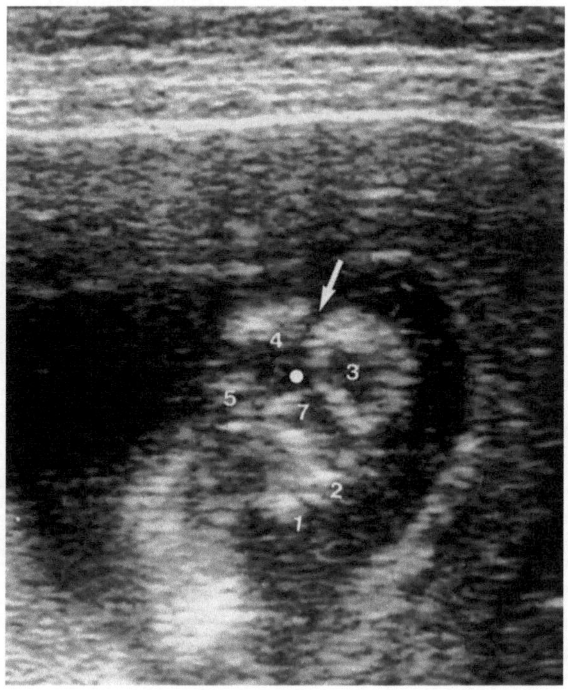

Fig. 4.6. Paramedian sagittal scan through a 12-week fetus. The arrow points to the coronal suture. The dot marks the sphenoid fontanelle

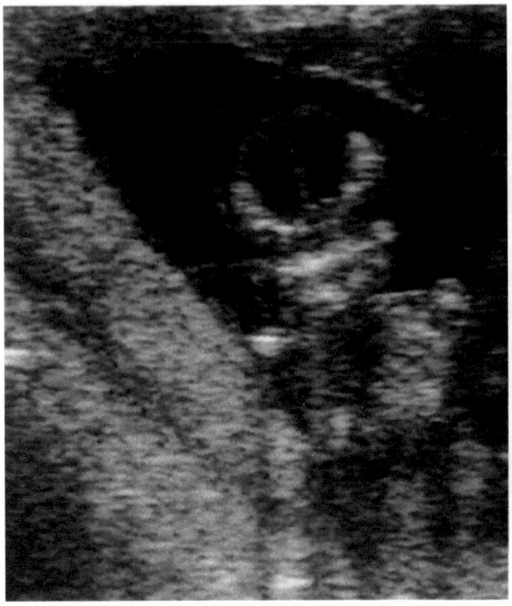

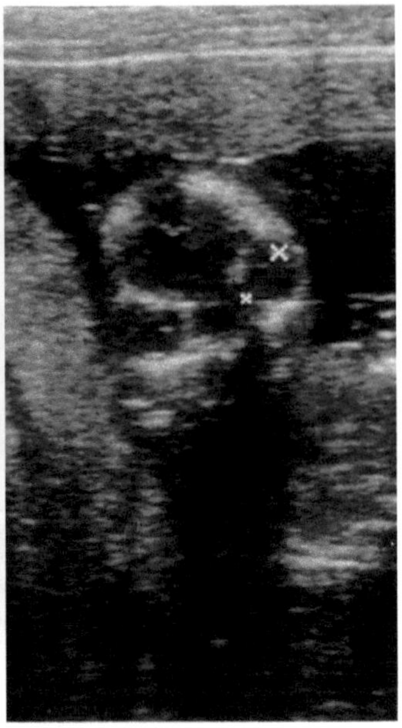

Fig. 4.7. (left) Frontal scan through a fetal skull at 13 weeks. The frontal bone is only partially ossified

Fig. 4.8. (right) Frontal scan through a skull at 17 weeks. Progressive ossification of the frontal bone allows delineation of the orbits, and the anterior fontanelle is visible

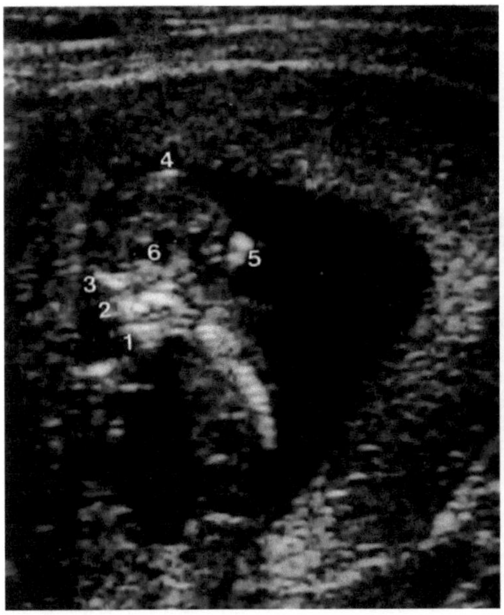

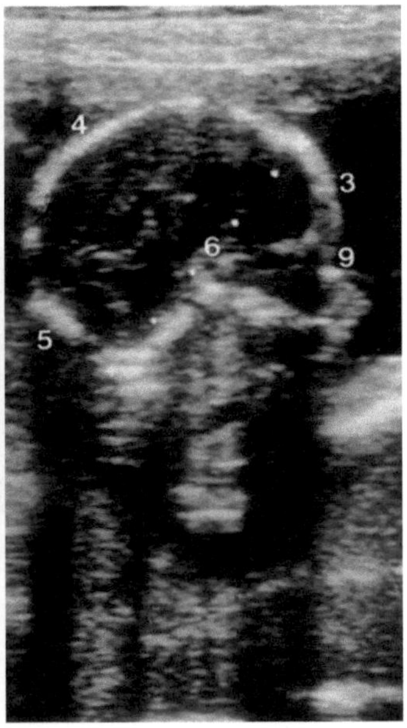

Fig. 4.9. (left) Paramedian sagittal scan through a fetal head at 13 weeks. The frontal bone, parietal bone, and occipital bone are partially ossified

Fig. 4.10. (right) Midsagittal scan through a fetal skull at 17 weeks. The cranial vault is completely ossified except for the fontanelles

Fig. 4.11. Exact midsagittal scan through a fetal skull at 15 weeks. The scan traverses the midline sutures and fontanelles, so there is little evidence of cranial ossification

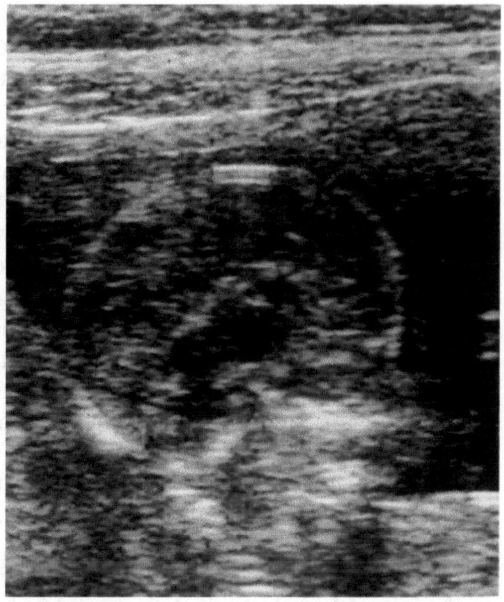

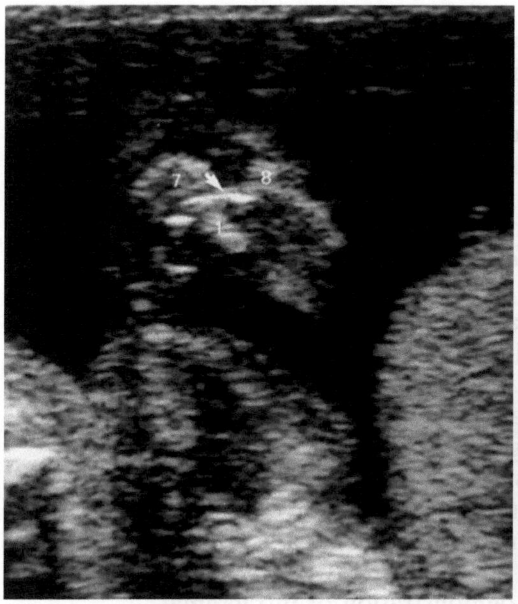

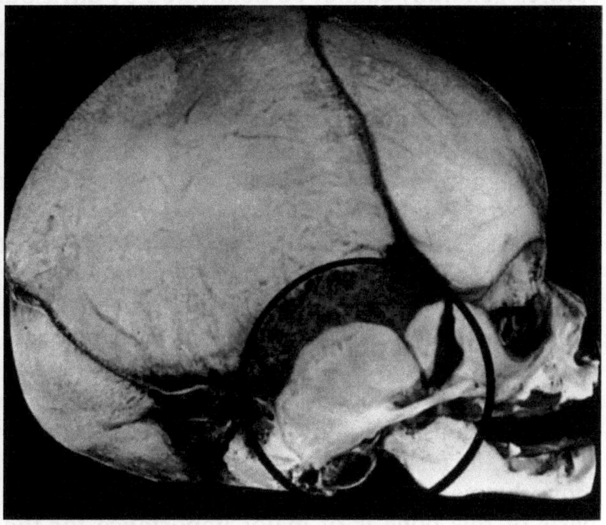

Fig. 4.12. a Tangential sagittal scan of a fetal head at 17 weeks delineating the temporal and zygomatic bones. The arrow indicates the zygomatic process of the temporal bone

Fig. 4.12. b The cranial bones demonstrated in Fig. 4.12 a are circled (19-week fetal skull)

The squama and zygomatic process of the temporal bone and the zygomatic bone itself can only be identified on a superficial tangential scan (Fig. 4.12 a). To aid orientation, the structures depicted on the sonogram are marked with a circle on a specimen from a 19-week fetus (Fig. 4.12 b).

The peculiar structure of the occipital bone merits some discussion. It is composed of six parts (Fig. 4.13): the two interparietal bones, the unpaired supraoccipital bone (these elements comprise the occipital squama), the paired exoccipitals, and the unpaired basilar occipital. The interparietal bones and supraoccipital bone fuse early. At 12 weeks the two interparietals are just visible on a posterior tangential scan (Fig. 4.14), and by 16 weeks they have completely fused into a rhomboid structure of homogeneous density (Fig. 4.15). The plane of these two scans is shown diagramatically in Fig. 4.16. The two exoccipitals and the unpaired basilar occipital still appear as separate structures on transverse scans in the third trimester (Fig. 4.17). They border the foramen magnum and apart from the tympanic ring (marked by arrows in Fig. 4.17) are the most prominent structures seen on transverse scans of the skull base. It is important to note that a sonolucent synchondrosis exists between the occipital squama and these three bony elements well into the third trimester, and that this synchondrosis is wide enough, together with the mastoid fontanelle, to serve as an acoustic window for scanning brain structures in the posterior fossa. In Fig. 4.18 this acoustic window is indicated on the skull of a 19-week fetus. On sagittal scans of the fetal head, the synchondrosis produces an apparent defect that must not be mistaken for a structural anomaly (Fig. 4.19).

Knowledge of the sutures and fontanelles of the fetal head is important 1) to avoid confusion with pathologic defects and 2) for utilizing these "gaps" as acoustic windows for examining the intracerebral anatomy.

Because of the curvature of the fetal cranium, significant portions of the cranial sutures generally cannot be delineated by ultrasound. An exception is the frontal suture, which is visible on superficial tangential scans of the forehead until about 21 weeks (Fig. 4.20). The coronal and lambdoid sutures produce apparent "defects" in the skull outline on transverse scans (Fig. 4.21), and the sagittal suture produces a similar effect on coronal scans up to 21 weeks (Fig. 4.22 b). The direct demonstration of fontanelles is possible only on occasional tangential scans before 20 weeks. Figure 4.23 shows a partially delineated anterior fontanelle, and Fig. 4.24 shows a tangential scan of the posterior fontanelle at 15 weeks. Both fontanelles appear as "structural defects" on frontal and sagittal scans (Fig. 4.22 a, 4.25 and 4.26) and provide important, physiologic acoustic windows for scanning the fetal as well as the neonatal brain (Babcock and Han 1981).

The sphenoid fontanelle and mastoid fontanelle create the impression of structural defects on both frontal and transverse views (Figs. 4.27– 4.29). As Fig. 4.29 shows, both of these fontanelles and the squamous suture that connects them form an ideal acoustic window for transverse scans of the brain. The corresponding image plane is shown schematically in Fig. 4.30. As mentioned earlier, the syndesmosis between the squamous part of the occipital bone and the two exoccipital parts of this bone is of major importance for the demonstration of intracranial detail. This acoustic window lies between the sonodense portions of the occipital squama (Fig. 4.31 a) and the three basal parts of the occipital bone (Fig. 4.31 c). This area, together with the mastoid fontanelle on each side, creates a large, uniform acoustic window that provides excellent access to the cerebellum (Fig. 4.31 b). The schematic drawing in Fig. 4.32 shows an overview of the fontanelles and sutures located about the lateral surface and base of the fetal skull.

The selective visualization of individual cranial bones is of minor importance in routine (Stage 1) scanning and is necessary only in rare cases. However, knowledge of the typical ultrasound features of individual bones can be very helpful in avoiding misinterpretations. To underscore the importance of an accurate evalua-

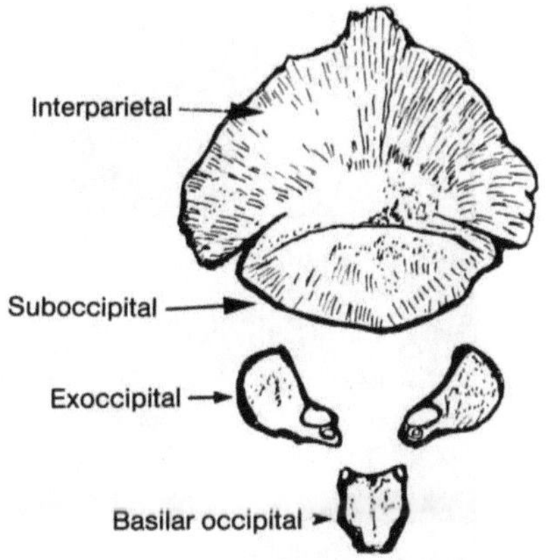

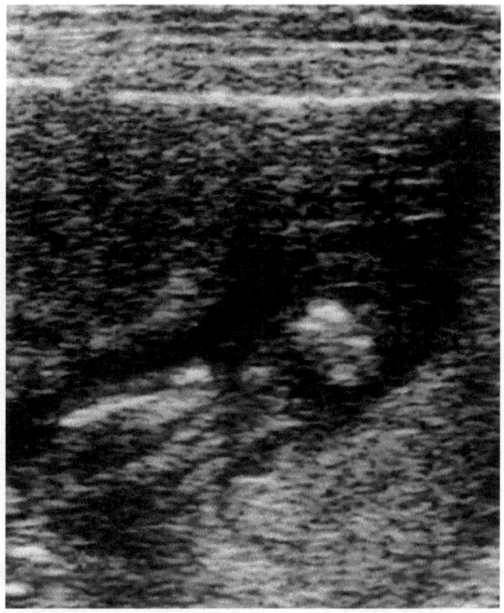

Fig. 4.13. (left) Development of the occipital bone (schematic)

Fig. 4.14. (right) Frontal scan in a 12-week fetus. The scan just cuts the occipital bone, and the two centers of the interparietal bone still appear as separate structures

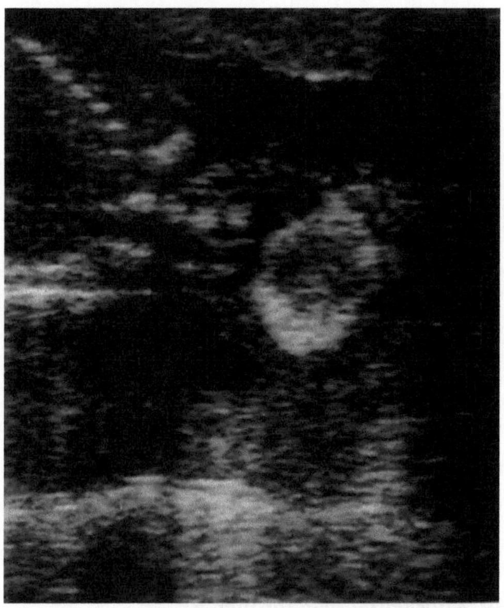

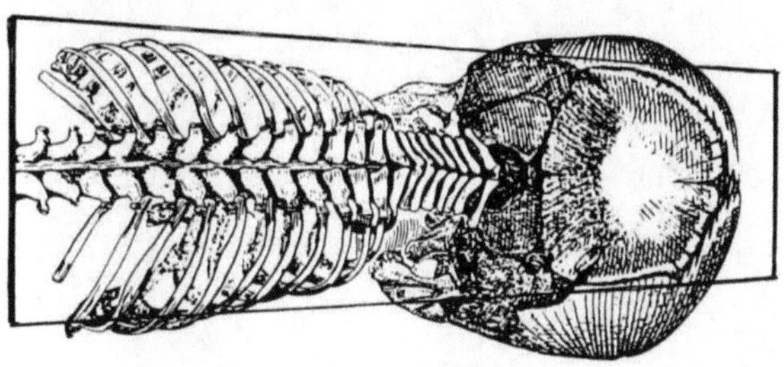

Fig. 4.15. (left) By 16 weeks the occipital squama presents as a single bone

Fig. 4.16. (right) Schematic diagram of the scan plane in Figs. 4.14 and 4.15

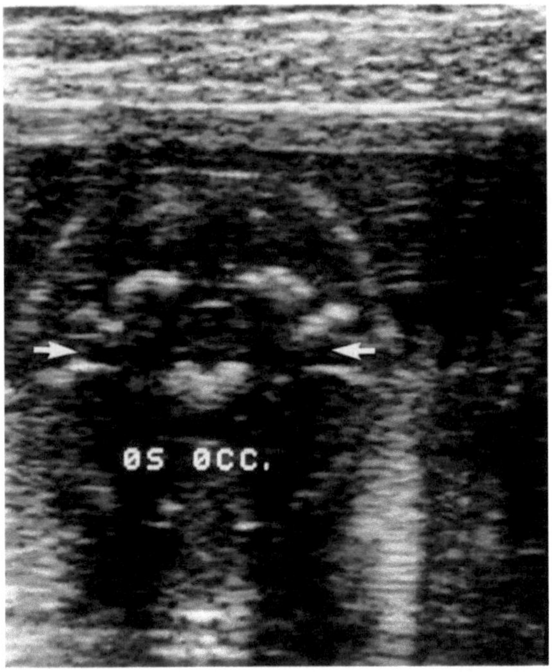

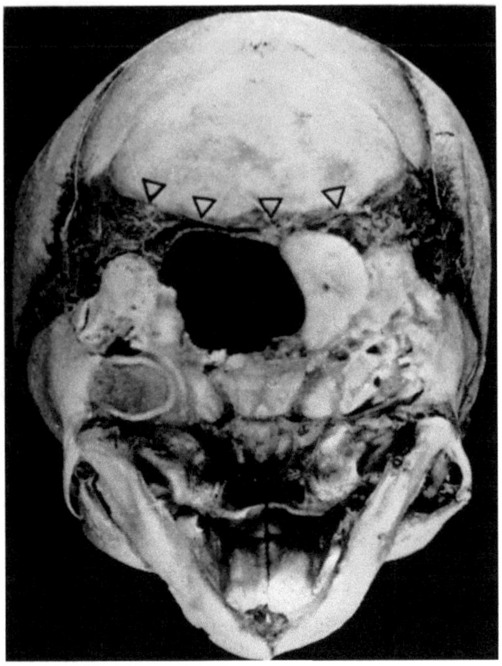

Fig. 4.17. (left) Scan through the skull base. The basal parts of the occipital bone appear as separate structures. *Arrows* mark the tympanic ring of the temporal bone

Fig. 4.18. (right) Fetal skull base at 19 weeks. The *arrows* mark the syndesmosis between the basal parts of the occipital bone and the occipital squama (acoustic window for scanning brain structures in the posterior fossa)

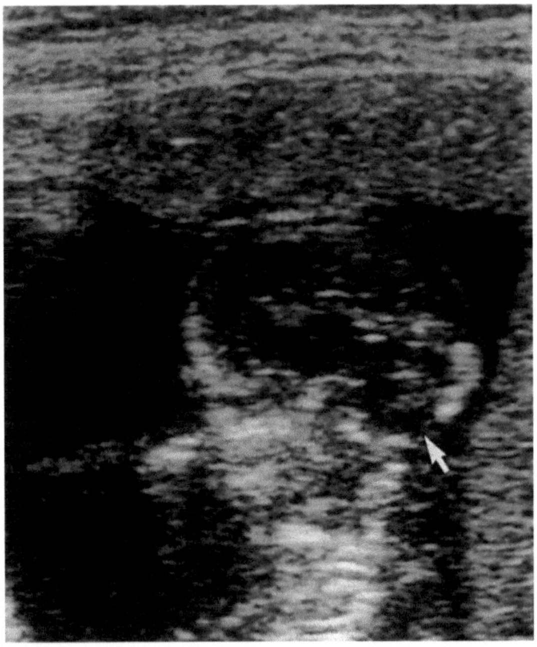

Fig. 4.19. Midsagittal scan through a fetal head at 13 weeks. The syndesmosis *(arrow)* produces a defect in the cranial contour

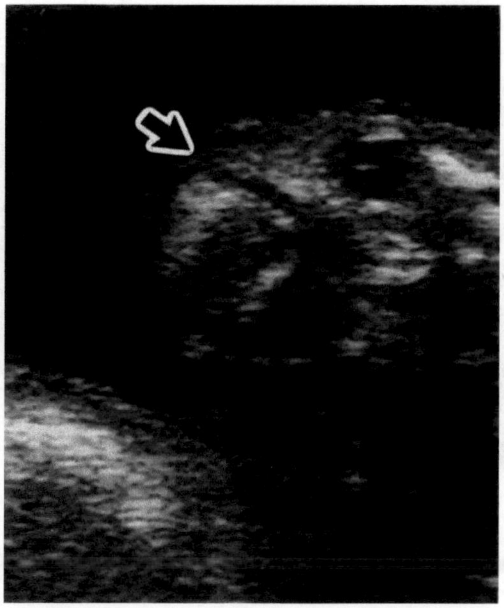

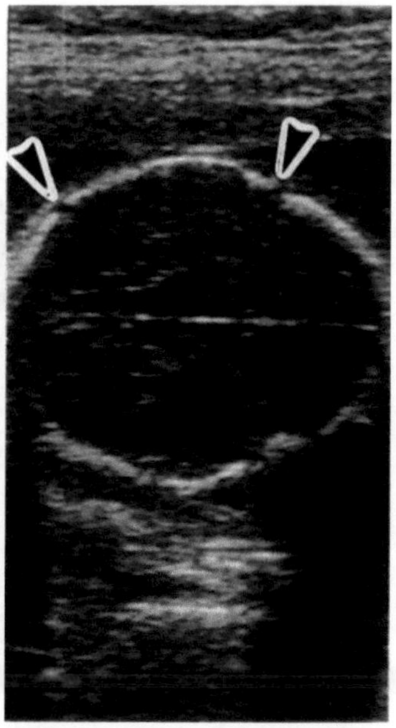

Fig. 4.20. (left) Tangential scan through the forehead of a 20-week fetus. The *arrow* indicates the frontal suture

Fig. 4.21. (right) Transverse scan through a skull at 20 weeks. The *arrows* mark the "gaps" formed by the coronal and lambdoid sutures

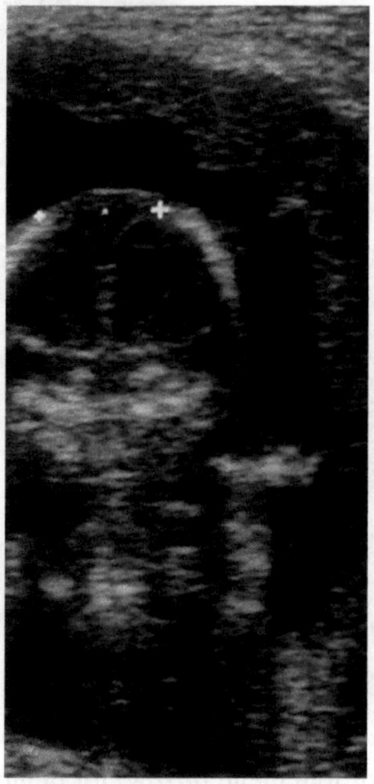

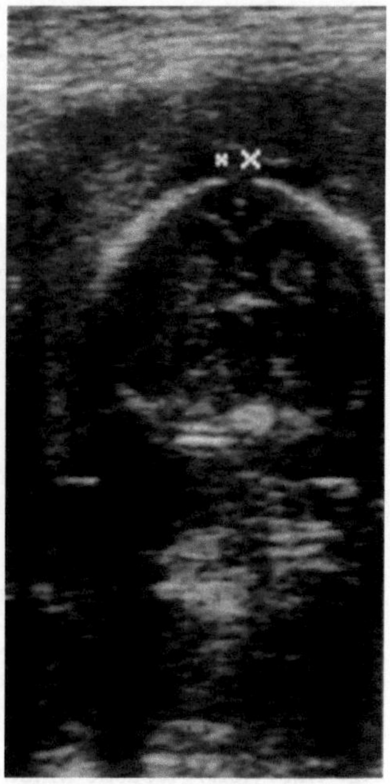

Fig. 4.22. a Frontal scan through a fetal skull at 17 weeks. The anterior fontanelle is visible in the cranial vault

Fig. 4.22. b Frontal scan through the same skull. On this more posterior scan the sagittal suture forms a small defect between the parietal bones

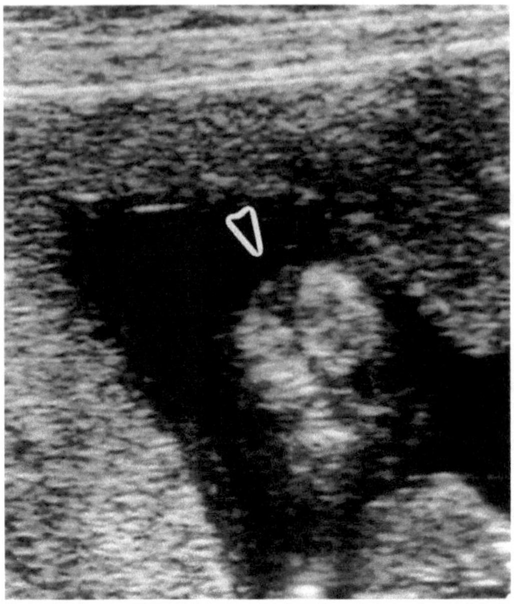

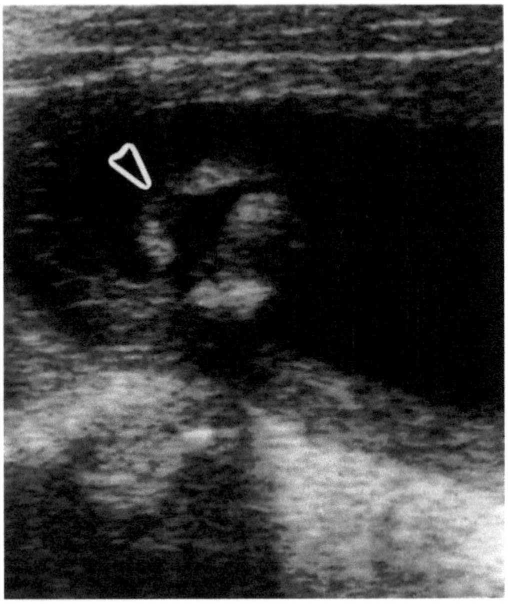

Fig. 4.23. (left) Tangential scan through the anterior fontanelle *(arrow)* in a 13-week fetus

Fig. 4.24. (right) Tangential scan through the posterior fontanelle *(arrow)* in a 15-week fetus

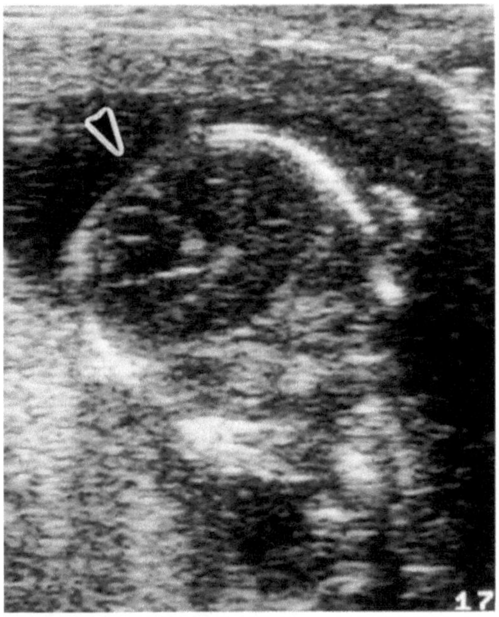

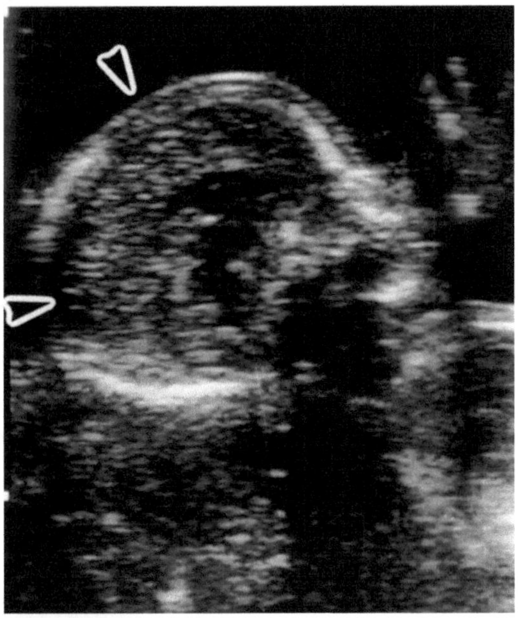

Fig. 4.25. (left) Acoustic window of the anterior fontanelle *(arrow)* for demonstrating the lateral ventricles in coronal section

Fig. 4.26. (right) Acoustic window of the anterior fontanelle *(arrow)* and posterior fontanelle *(arrow)* in sagittal section. The scan cuts the falx cerebri exactly in the midline. Scans performed through the acoustic windows give excellent views of intracranial structures

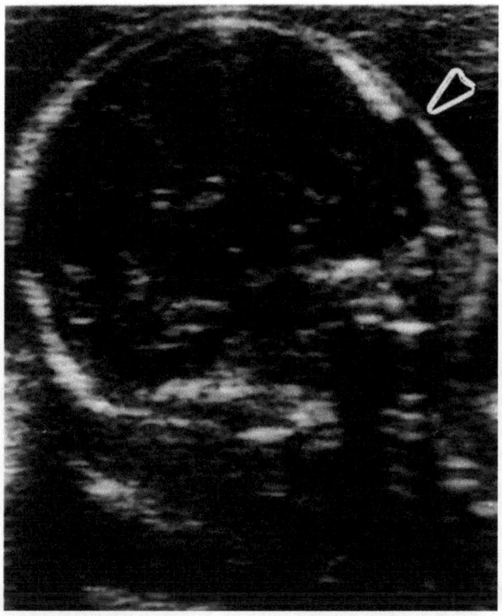

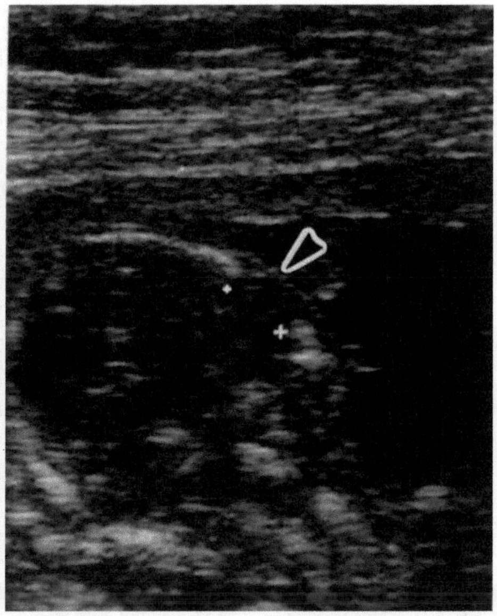

Fig. 4.27. (left) Frontal scan through a skull at 20 weeks. The *arrow* marks the "defect" of the sphenoid fontanelle

Fig. 4.28. (right) The „defect" of the mastoid fontanelle is indicated

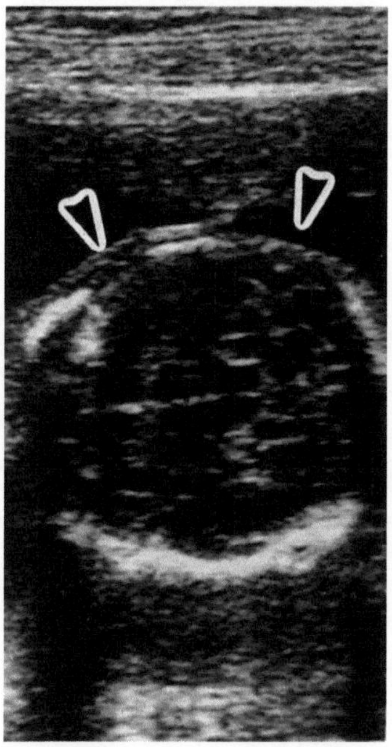

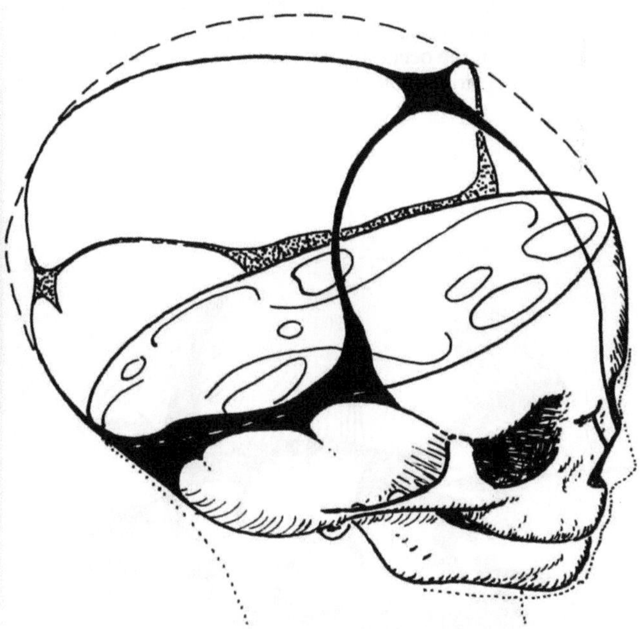

Fig. 4.29. (left) Transverse scan through the acoustic windows of the sphenoid and mastoid fontanelles just above the level of the skull base. Intracerebral structures can be imaged with exceptional clarity through these acoustic windows

Fig. 4.30. (right) Diagram of the scan plane obtained through the lateral acoustic windows (mastoid fontanelle, sphenoid fontanelle)

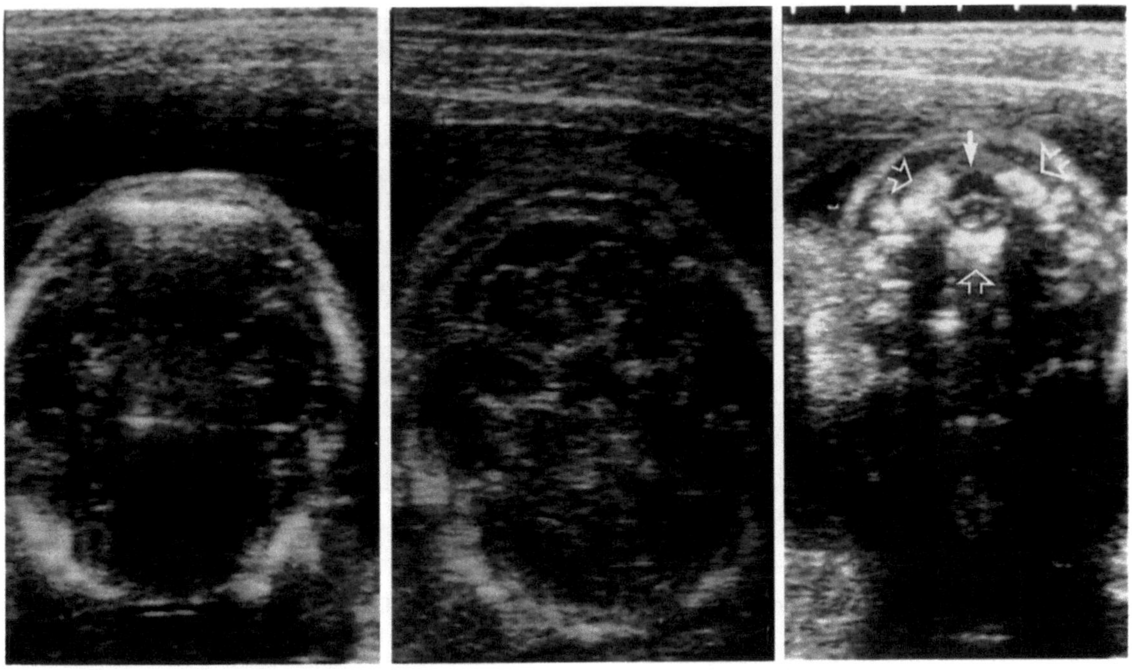

a b c

Fig. 4.31. **a** Transverse scan of a skull in the occipitoposterior position. The scan plane cuts the occipital squama, so brain structures are not visualized. **b** The transducer has been moved caudally to scan through the acoustic window of the syndesmosis between the occipital squama and basal part of the occipital bone. The cerebellum and cisterna magna are very clearly demonstrated. **c** The basal portions of the occipital bone are visualized by moving the transducer farther caudally. The spinal cord is clearly visible in the foramen magnum

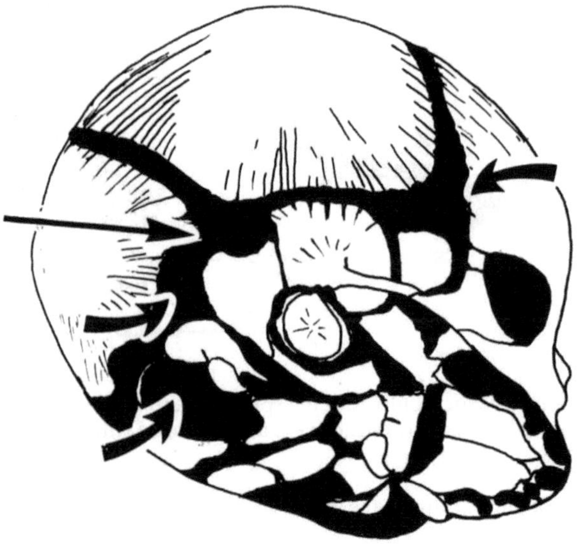

Fig. 4.32. Diagram showing the location of the lateral and posterior acoustic windows

Table 4.2. Syndromes with predominant involvement of the cranial bones (Schmid 1973)

a) Dysmorphic syndromes	b) Cranial syndromes
Apert syndrome (acrocephalosyndactyly)	Ankyloglossia superior syndrome
Crouzon syndrome (craniofacial dysostosis)	Capdepont syndrome
Pseudo-Crouzon syndrome	Cherubism syndrome
Holtermüller-Wiedemann syndrome	Christ-Siemens-Touraine syndrome
Scheuthauer-Marie-Sainton syndrome (cleidocranial dysostosis)	Costen syndrome
Waardenburg syndrome (II) (dyscephalosyndactyly)	Crouzon syndrome
Ulrich-Feichtiger syndrome (dyscraniopygophalangia)	Franceschetti syndrome
Freeman-Sheldon syndrome (craniocarpotarsal dystrophy)	Garcin syndrome
De Lange syndrome	Gregg syndrome
Oculo-dento-digital dysplasia (oculodentodigital syndrome, Meyer-Schwickerath-Weyers)	Holtermüller-Wiedemann syndrome
Klippel-Feldstein syndrome	Jacod syndrome
Oculovertebral syndrome (Weyers)	Kurz syndrome
Gruber syndrome (splanchnocystic dysencephaly)	Maxillofacial syndrome
Rubinstein syndrome	Maxillonasal syndrome
Franceschetti syndrome (I) (manibulofacial dysostosis)	Nager-de Reynier syndrome
Weyers syndrome (acrofacial dysostosis)	de Sanctis Cacchione syndrome
Hanhart syndrome (II)	Ullrich-Fremerey-Dohna syndrome
Ullrich-Fremerey-Dohna syndrome (dyscraniodysopia)	
Hallermann syndrome (mandibulo-oculo-facial dysmorphia, probably an oligosymptomatic variation of de Lange syndrome)	
Mandibular dysostosis	
Otocephaly	
Gregg syndrome	
Dzierzynsky syndrome	
Achondroplasia	
Enchondral dysostosis	

tion of the bony cranium and facial skeleton in Stage 2 and 3 examinations, and especially in patients referred for exclusion of fetal anomalies, syndromes that predominantly involve the cranial bones are listed in Tables 4.2a and 4.2b along with typical cranial syndromes (Schmid 1973). The latter source should be consulted for more detailed information on the various syndromes.

4.3 Fetal Brain Anatomy

Until a few years ago, the major purpose of visualizing the fetal brain by ultrasound was to demonstrate the midline echo for the measurement of the BPD. More detailed studies of the fetal brain were not possible because of the limited resolution of available equipment. Important details of fetal intracerebral anatomy have been published by Denkhaus and Winsberg (1979), Johnson and Rumack (1980), Hadlock et al. (1981 a), Hobbins et al. (1983), and Hansmann et al. (1985). To our knowledge, no comprehensive study has previously been published which compared sonograms taken in standardized planes with actual anatomic sections.

Table 4.3. The principal structures of the fetal brain that can be demonstrated by ultrasound (numbers correspond to key numbers in the figures)

1	Frontal horn (lat. ventr.)	16	Lamina tecti
2	Cavum septi pellucidi	17	Pons
3	Thalamus	18	Fornix
4	Cisterna venae magnae cerebri	19	Cisterna magna
5	Occipital horn (lat. ventr.)	20	Corpus callosum
6	Insula	21	Inferior horn (lat. ventr.)
7	Third ventricle	22	Capsula interna
8	Cerebral peduncle	23	Lentiform nucleus
9	Cerebral aqueduct	24	Caudate nucleus
10	Cerebellum	25	Hippocampus
11	Choroid plexus	26	Sphenoid bone
12	Brain mantle	27	Orbit
13	Interpeduncular cistern	28	Occipital lobe
14	Basilar artery	29	Cingulate sulcus
15	Falx cerebri	30	Parieto-occipital sulcus

As mentioned in Sect. 2.1 on Frozen Section Technique, we prepared gross sections from the brains of 122 spontaneously aborted fetuses. Sections from 38 of these fetuses were suitable for use; the rest were unacceptable because of hemorrhage or tissue damage. We prepared transverse sections in 20 fetuses, analogous to the views obtained in neonatal ultrasound (Babcock and Han 1981). Using cranial surface landmarks, we located the canthomeatal line and performed the sections at an angle of about 10° to that line. In 10 other fetuses we obtained frontal sections at right angles to that plane, and in 8 fetuses we performed sagittal and parasagittal sections. Intracranial structures that were consistently seen on routine scans but could not be immediately identified were demonstrated in vitro in freshly aborted fetuses and marked with a small methylene blue depot injected under ultrasound guidance so that the structure could later be located on frozen sections and anatomically identified.

Transverse sections were used to establish the true size of the lateral ventricles. This was done after approximately four weeks' freezer storage to achieve a clear demarcation of surface features. The ventricles were dissected free, and their dimensions established by placing the sections on millimeter grid paper (see Sect. 2.1).

In this chapter we use numbers to label the main intracerebral structures that appear on sonograms and in frozen sections. The structures and their numeric labels are listed in Table 4.3.

For didactic reasons we base the discussion of fetal intracranial anatomy not on individual brain structures but on the standard planes of section that are of diagnostic importance.

4.3.1 Transverse Sections

It is almost always possible to obtain transverse scans of the fetal head by appropriate transducer placement on the maternal abdomen, regardless of the fetal presentation and position. In theory, the fetal head can be scanned axially at any number of levels, but in practice there are six transverse planes that exhibit characteristic sonoanatomic structures and can be reproducibly obtained. These standard planes are shown diagramatically in Fig. 4.33. They are also marked on median and paramedian sagittal anatomic sections (Fig. 4.34) to illustrate more clearly the relative positions of the planes and the structures they traverse.

Transverse plane 1. Figure 4.35 a–c shows the sonogram, frozen section, and topographic drawing of transverse plane 1 in a 17-week fetus. The uninterrupted midline echo is formed primarily by the falx cerebri. The demarcation between the lateral ventricles and brain mantle (wall of the hemisphere) is less distinct on the sonogram than in the anatomic section. On the sonogram we see marked differences in the appearance of the structures of the near-field cerebral hemisphere (the hemisphere closer to the transducer) and the far-field hemisphere. The former appears more echo-dense, while the latter appears relatively "echo-free".

In experimental water-bath studies of aborted fetuses, we were able to show that this phenomenon is based both on the distance between the fetal head and transducer and on the overall gain setting of the equipment (Staudach and Lassmann 1984). The sonodense area in the ventricle close to the transducer is a reverberation artifact; the far ventricle presents an accurate picture of the anatomic structures of the hemisphere.

The most prominent intraventricular structure is the choroid plexus, which is highly echogenic and fills the lateral ventricle almost completely except for the frontal horn. No other important intracerebral features can be recognized on the sonogram or frozen section at this level.

The major purpose of a scan at the level of transverse plane 1 is evaluation of the lateral ventricles. Kiel (1977) has described the main radiologic features of the developing lateral ventricles, but ultrasound studies have dealt with this question only from the aspect of recognizing ventricular enlargement as a means of diagnosing hydrocephalus. In the literature, the earliest sign of hydrocephalus is said to be dilatation of the occipital horn, followed by the frontal horn of the lateral ventricles (Weisbert et al. 1978; Hobbins et al. 1979; Johnson et al. 1980a; Fiske et al. 1981; Hansmann 1981; Hansmann et al. 1985). Because dilatation of the ventricles occurs before the fetal head enlarges, the early diagnosis of hydrocephalus in the second trimester is directly related to the demonstration of the ventricles (Hansmann 1981).

Campbell (1979) introduced the ventricular hemispheric ratio (VHR) as a means of evaluating this expansion. The VHR represents the distance of the lateral wall of the lateral ventricle from the midline as a ratio over the maximum hemispheric width. This ratio should be less than 0.5 after 17 weeks. The distance from the lateral ventricle to the inner table of the skull has been introduced as another criterion (Vintzileos et al. 1983). Johnson et al. (1980a) determined the lateral ventricular width directly by measuring the distance from the middle of the midline echo to the most prominent echo of the lateral wall of the lateral ventricle in the area where the ventricular wall is parallel to the midline echo and the hemispheric width is maximal. These authors also determined the hemispheric dimension by measuring the distance from the midline echo to the inner table of the skull. They used both measurements to construct a ratio, which, however, shows an extremely wide 2 s range throughout gestation. All these measurements disregard the actual width of the cerebral mantle medial to the inner wall of the lateral ventricle.

Our own investigations in fetal frozen brain sections cast doubt on the general usefulness of such tables. Biometry in these planes appears to have very little value from 12 to 15 weeks. Even in the normal brain, at this gestational age the wall of

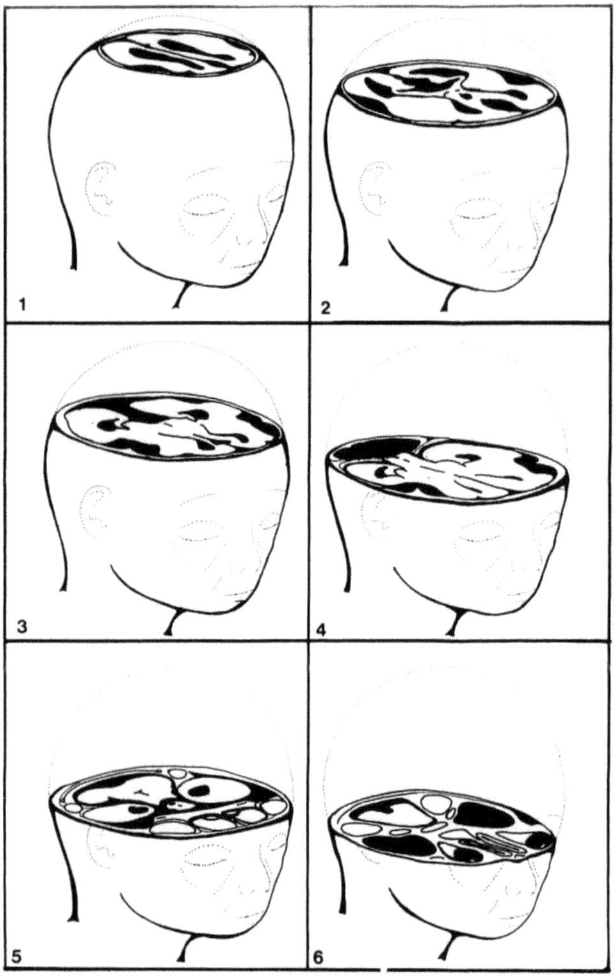

Fig. 4.33. (left) Diagrams showing the anatomical relationship of the six major transverse planes of section of the fetal head

Fig. 4.34. (below) The levels of the major transverse planes in relation to midsagittal and paramedian sagittal sections through a fetal head (21 weeks)

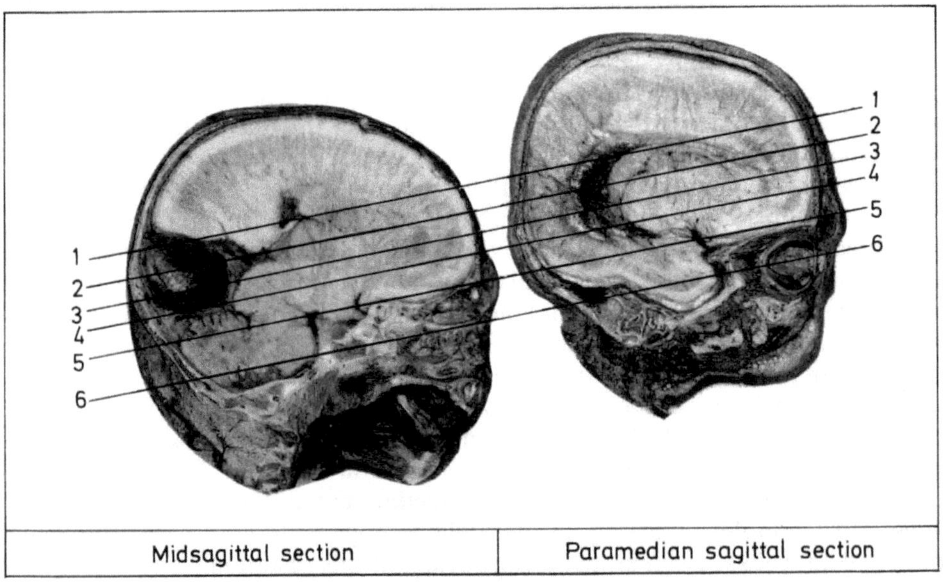

the hemisphere is very narrow in relation to the ventricles. The choroid plexus fills the lateral ventricles almost completely, except for the frontal horns, and is highly echogenic (Fig. 4.36 a, b). This probably results from the irregular surface of the plexus and its vascularity. The predominance of the choroid plexus on sonograms decreases as gestation progresses. Studies by Crade et al. (1981) showed that the size ratio of the plexus to the lateral hemisphere declines steadily from week 12 to week 21. In the frozen sections, the plexus could hardly be demonstrated as a separate structure within the lateral ventricles after 24 weeks. The frozen section in Fig. 4.37 shows the relative size of the lateral ventricles and brain mantle at 17 weeks, with the contents of the left lateral ventricle removed (dimensions shown in millimeters).

In five fetuses of comparable gestational age, we found identical ventricular widths in this reference plane. The maximum ventricular width was 3 mm anteriorly and centrally, and 5 mm posteriorly. The medial brain mantle between the lateral ventricle and falx showed a constant thickness of 3 mm. Comparing these values with the measurements of Johnson et al. (1980a), we find that these authors report a mean lateral ventricular width greater than 8 mm at a comparable stage of gestation. Even allowing for the fact that this measurement disregards the true width of the medial brain mantle, this discrepancy is too large. Figure 4.38 shows the ventricular width in the same plane in a section from a 19-week fetus. We see that the dissected left ventricle has expanded to 5 mm frontally and centrally, and to 6 mm occipitally. The slight obliquity of this section causes the occipital horn of the right ventricle to appear enlarged relative to the left side. Slight shifts of the scan plane are unavoidable in ultrasound examinations, and this fact alone casts doubt on the general validity of this measurement technique. Only 2 fetuses were available for ventricular width measurements at 23 weeks. Isolated dissection of the lateral ventricles was not possible in either of these cases, so the widths could not be accurately established. Figure 4.39 shows a frozen section from a 23-week fetus. Gross examination of the lateral ventricle on plane 1 showed a maximum ventricular width of 4 mm. The lateral brain mantle had expanded from 8 mm at 19 weeks to now 25 mm. The medial thickness of the brain mantle ranged from 6 to 9 mm, depending on the site of the measurement. Although the case numers are too small for statistical analysis, the measurements in the frozen sections demonstrate the problems that are associated with sonographic measurements in this area.

Pediatric studies of ventricular width are available after 26 weeks. Although these studies employ a different plane than in antenatal measurements, the maximum values rarely exceed 3 mm (Fiske et al. 1981; Perry et al. 1985).

Examiners should be cautious about diagnosing hydrocephalus before 23 weeks. The false-positive rate for referrals in our series was highest in patients who were referred because of suspected fetal hydrocephalus (see Sect. 2.2). In all cases of hydrocephalus confirmed before 23 weeks, we found obvious evidence of intracerebral pathology on multiple planes. The paramedian sagittal scan, not yet described in the literature, appears to be particularly suitable for measurements of lateral ventricular size (see p. 75).

Transverse plane 2. This scan is obtained by moving the transducer several millimeters caudally from plane 1, keeping parallel to that plane. Plane 2 generally is skipped during routine scanning, but as it can be reproducibly obtained, its major features are illustrated in Fig. 4.40 a–c. Figure 4.40 a shows the sonogram from a 17-week fetus, and Fig. 4.40 b shows the corresponding anatomic section. Note the

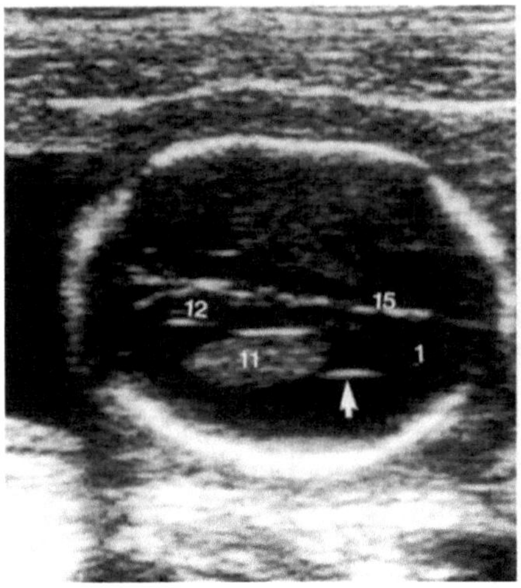

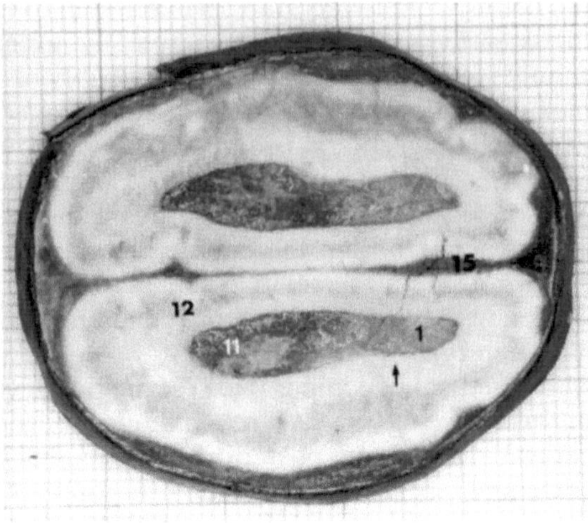

Fig. 4.35 a–c. (top and center) Transverse plane 1. **a** Sonogram. **b** Frozen section. **c** Diagram. *1* Frontal horn (lat. ventr.); *11* choroid plexus; *12* brain mantle; *15* falx cerebri

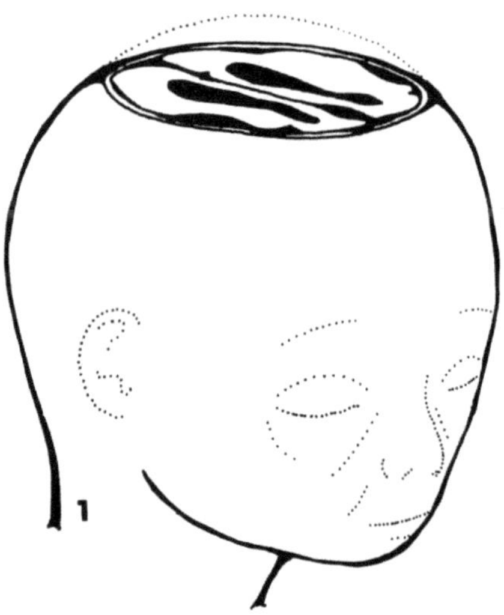

Fig. 4.36 a, b. (below) Transverse plane 1 at 14 weeks. The brain mantle is thin, the ventricles are large and completely filled by choroid plexus. **b** Analogous frozen section (14 weeks)

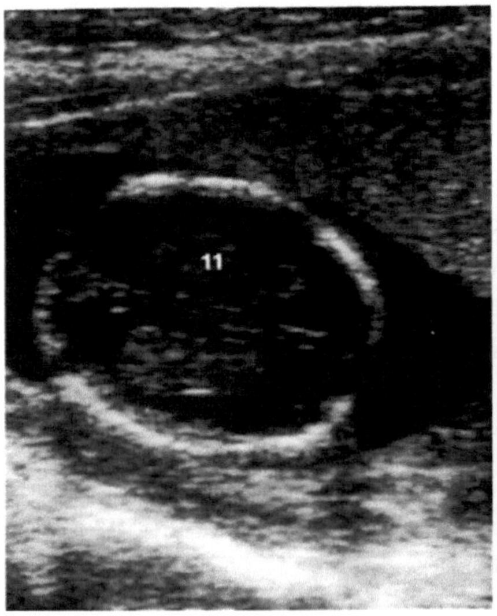

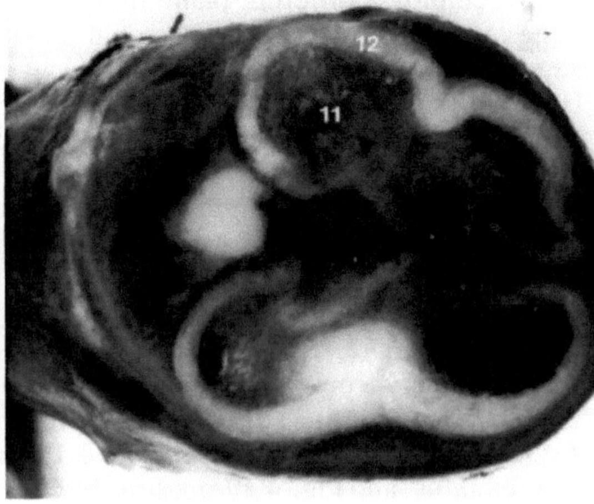

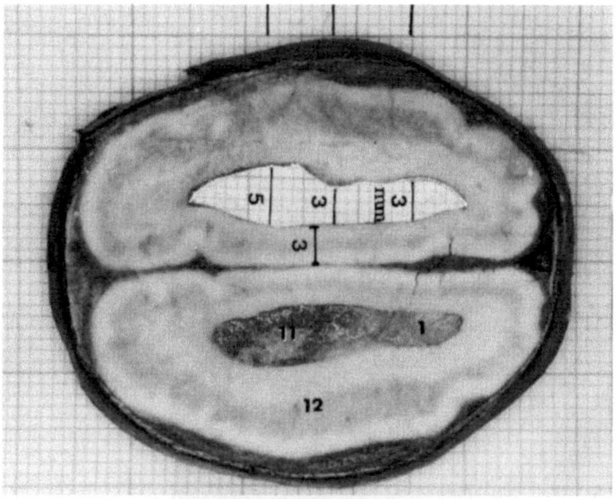

Fig. 4.37. Frozen section from a 17-week fetus. Transverse plane 1, with the left ventricle dissected free to show its frontal, central, and occipital dimensions (in mm)

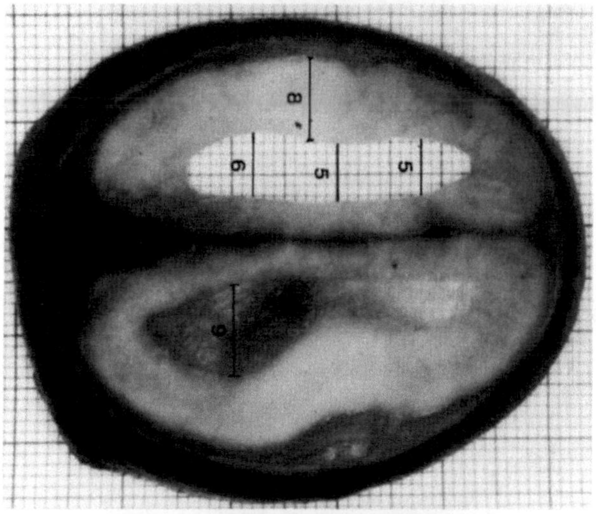

Fig. 4.38. Frozen section from a 19-week fetus. The left lateral ventricle has been dissected free to show its dimensions relative to the brain mantle. The slight obliquity of the section causes an apparent widening of the right occipital horn

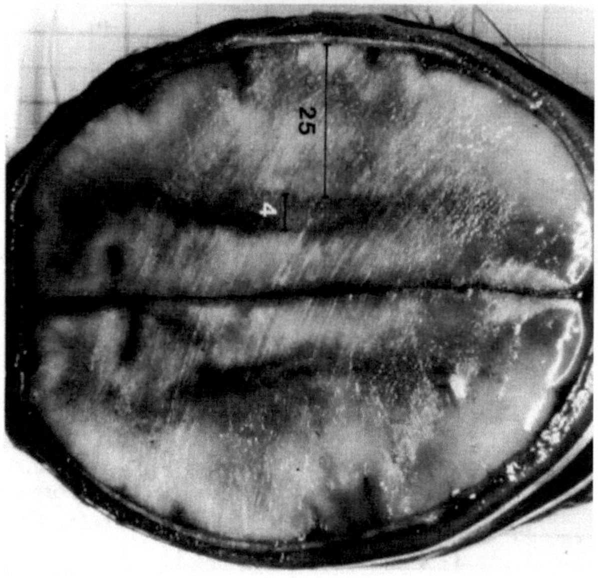

Fig. 4.39. Frozen section at the level of transverse plane 1 from a 23-week fetus. The thickness of the brain mantle has increased markedly relative to the ventricles

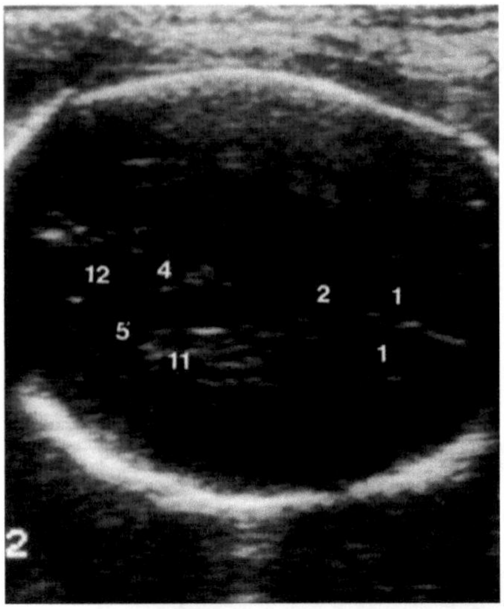

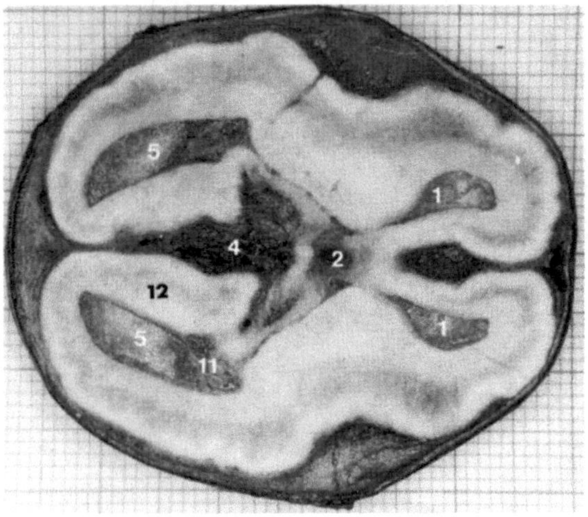

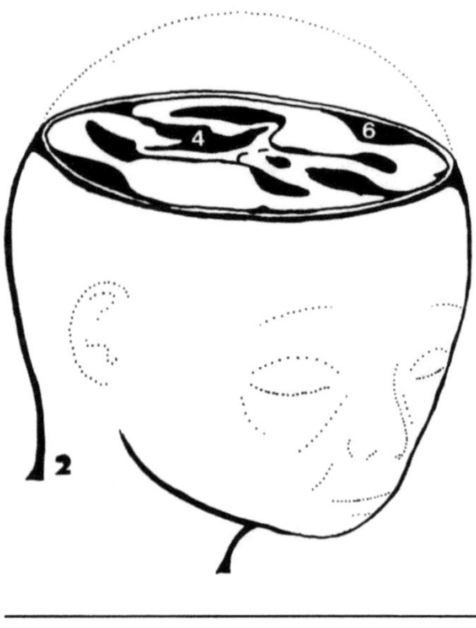

Fig. 4.40 a–c. Transverse plane 2. **a** Sonogram. **b** Frozen section. **c** Diagram. *1* Frontal horn (lat. ventr.); *2* cavum septi pellucidi; *4* cisterna venae magnae cerebri; *5* occipital horn (lat. ventr.); *6* insula; *11* choroid plexus; 12 brain mantle

central interruption of the midline echo and the H-shaped configuration of the ventricles. The central echo-free area adjacent to the frontal horns is the cavum septi pellucidi. The thalami are not visible at this level. The occipital horns are separated by an echogenic area representing the cisterna venae magnae cerebri.

Transverse plane 3. This plane is of key importance, as it is the standard reference plane for biometry of the fetal head. Figure 4.41 a–c shows the sonogram, anatomic section, and topographic drawing for this plane in a 17-week fetus. A continuous midline echo is no longer demonstrated at this level. The falx appears only as a dividing line between the frontal horns. The faint midline structure directly behind the cavum septi pellucidi represents the slit-like third ventricle, which is flanked by semicircular, moderately echogenic areas representing the thalami. Directly behind the thalami is the cisterna venae magnae cerebri, which appears triangular in this plane. This cistern is bordered on both sides by the occipital horns, which contain the choroid plexus.

A common source of misinterpretation in this plane is the apparent paucity of echoes in the far-field hemisphere compared to the near hemisphere. This impression is reinforced by the presence of the insula. Jeanty et al. (1984b) note that the insula is commonly identified as the sylvian fissure, which is ontogenically incorrect and results from the comparison of sonographic features with anatomic sections from adults (Johnson et al. 1980; Grant et al. 1981; Hadlock et al. 1981). The insula is clearly visible in the anatomic section in Fig. 4.41 b; at this stage it is not yet narrowed or covered by the opercula of the frontal and temporal lobes. The insula is still freely accessible in the frontal section at 21 weeks (Figs. 4.42 a, b and 4.43). After 24 weeks the insula becomes increasingly covered by the frontal and temporal lobe, and it is not until this time that the true sylvian fissure begins to appear (Dorovini-Zis and Dolman 1977). The maximum distance between the surface of the insula and the inner table of the skull was 6 mm at 17 and 19 weeks (Figs. 4.44 and 4.45) and 9 mm at 21 weeks (in the frontal section in Fig. 4.42 b). This section is readily identified on transverse as well as frontal scans, as it is clearly marked by the pulsating middle cerebral artery (Fig. 4.46). The third ventricle presents only as a linear midline echo whose dimension cannot be measured, regardless of gestational age (Fig. 4.46). We were unable to demonstrate the third ventricle as an isolated structure in any of our frozen sections.

The transverse plane 3 is especially important for biometry of the fetal head. It is the plane in which the BPD, OFD, and head circumference are determined.

The BPD has been an important biometric parameter for years, and the measurement procedure is well known (Donald and Brown 1961; Willocks 1963; Willocks et al. 1964; Campbell 1968; Hofmann and Holländer 1968; Holländer 1972, 1975, 1984; Hansmann et al. 1985). Measurement of the OFD was described by Hansmann (1976), Levy and Erbsman (1975), Schillinger et al. (1976), and by Hansmann et al. (1985). Holländer (1984) states that he abandoned measurement of the OFD because he could not find a criterion for determining whether the scan obtained actually represented the occipitofrontal plane. In our studies of spontaneously aborted fetuses, we tried to find intracerebral landmarks that would accurately define the plane for measuring the OFD. We first measured the maximum biparietal and fronto-occipital diameters on fetal heads directly with a caliper before freezing and inserted needles to mark the points of maximal separation (Fig. 4.47 a, b). Later the needles served as guides for sectioning of the specimens. We found that the structural features of the section defined by the needles corre-

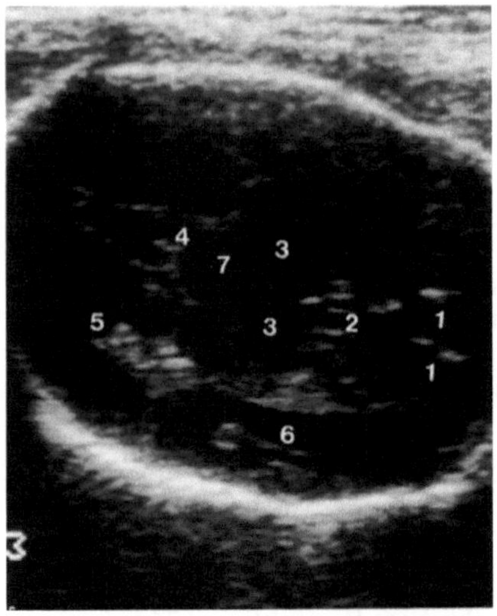

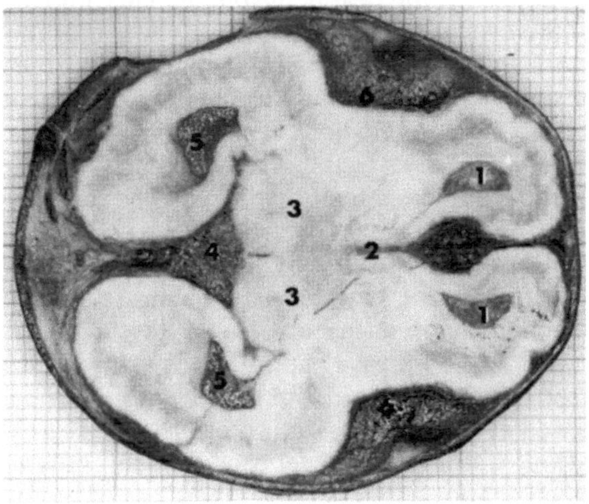

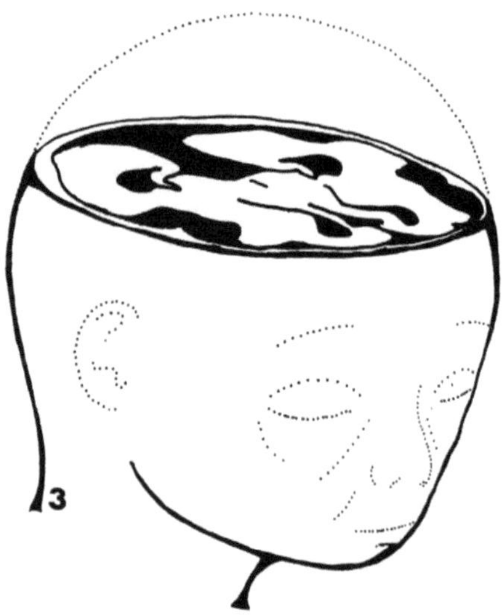

Fig. 4.41 a–c. Transverse plane 3. **a** Sonogram. **b** Frozen section. **c** Diagram. *1* Frontal horn (lat. ventr.); *2* cavum septi pellucidi; *3* thalamus; *4* cisterna venae magnae cerebri; *5* occipital horn; *6* insula; *7* third ventricle

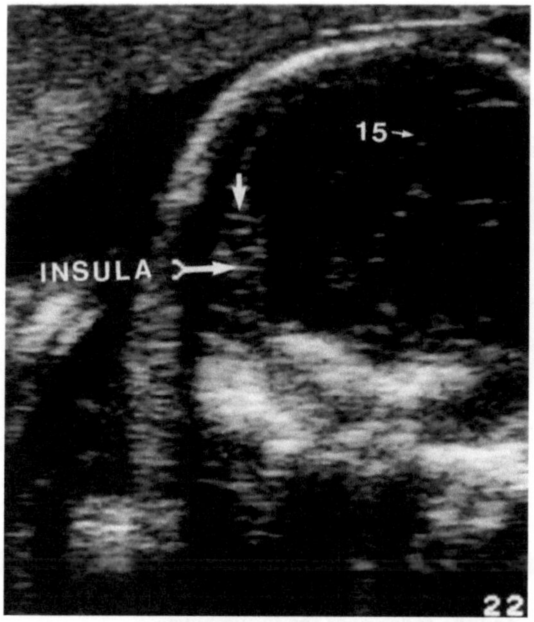

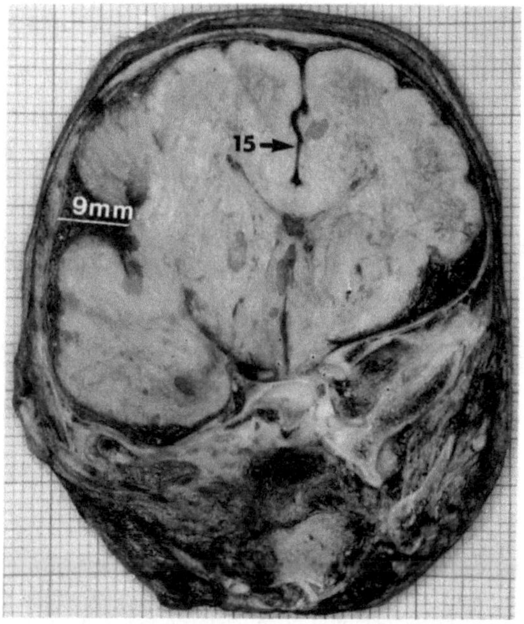

Fig. 4.42. a Frontal scan through a skull at 21 weeks. The *arrows* point to the insula and the middle cerebral artery

Fig. 4.42. b Analogous frozen section

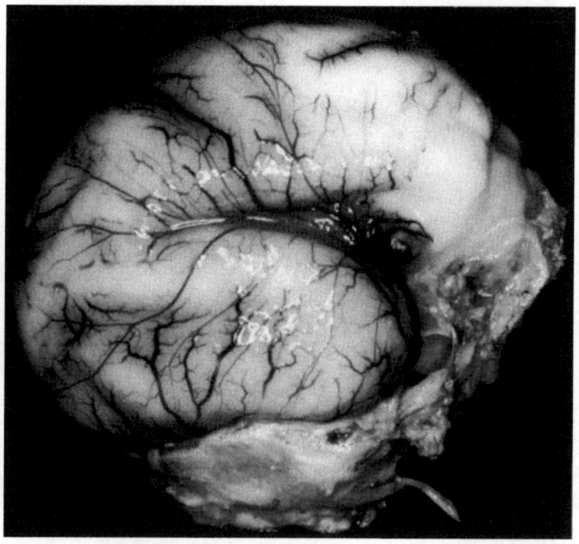

Fig. 4.43. View of a fetal brain at 22 weeks. The insula is freely visible and not yet covered by the temporal and occipital lobes

sponded without exception to "transverse plane 3", regardless of gestational age. Ultrasound biometry was performed using a tissue velocity of 1540 m/s and placement of measuring points outside to outside.

This plane, then, is defined by the following intracerebral landmarks: the frontal horns and cavum septi pellucidi anteriorly, the thalami and third ventricle cen-

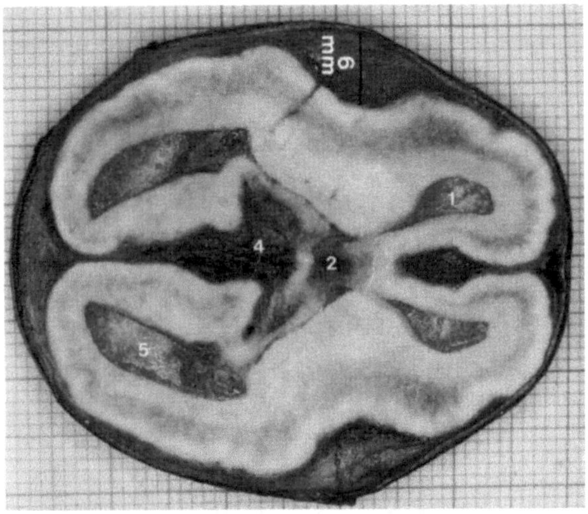

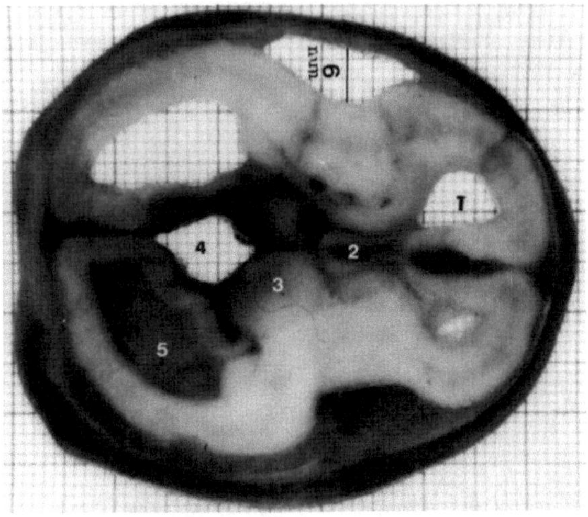

Fig. 4.44. (left) Transverse plane 2 through a fetal skull at 17 weeks. The surface of the insula is 6 mm from the cranial wall

Fig. 4.45. (right) Transverse plane 3 through a fetal skull at 19 weeks. In the left hemisphere the frontal horn, occipital horn, cisterna venae magnae cerebri, and the space over the insula have been dissected free

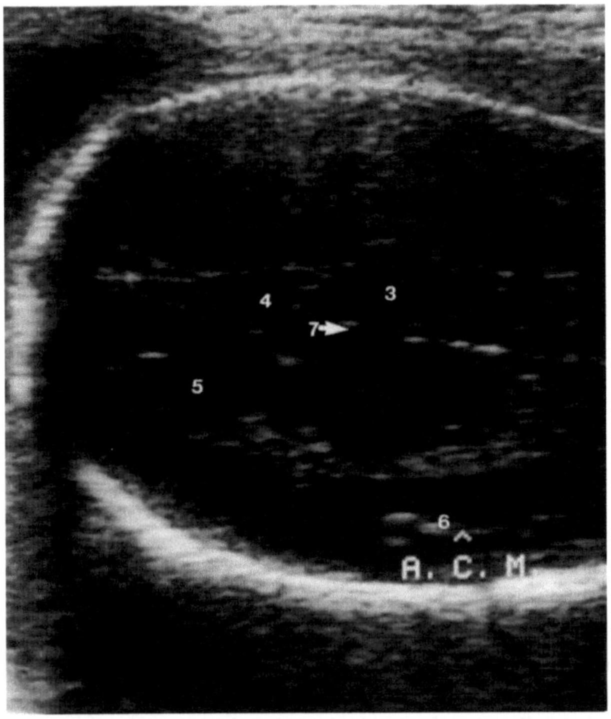

Fig. 4.46. Scan on transverse plane 3 at 23 weeks. The third ventricle *(arrow)* forms a slit-like space between the thalami. The middle cerebral artery (ACM) is visible in the area of the insula

trally, and the occipital horns and cisterna venae magnae cerebri posteriorly (Figs. 4.41 and 4.46). Figure 4.48 a–c shows how plane selection can affect the assessment of gestational age based on head parameters. All three scans show a symmetrical cross section of the fetal head, and all demonstrate a midline echo. The scan in Fig. 4.48 a corresponds to transverse plane 1. The BPD on that scan is 66 mm, the OFD is 86 mm, and the head circumference (HC) is 252 mm, which would correspond to 26 weeks according to a standard growth curve (Hansmann et al. 1985). Figure 4.48 c shows a scan obtained caudal to standard plane 3 (the cerebellum is visible posteriorly). The BPD at this level is 67 mm, the OFD is 88 mm, and the HC is 257 mm, corresponding to a gestational age of 27 weeks. Figure 4.48 b shows a scan performed at the optimum level for measuring the BPD and OFD according to our investigations. The HC in this plane is 269 mm, corresponding to 28 weeks. In the least favorable case, then, an error of plane selection can result in an error of two weeks when HC is used to determine gestational age.

Transverse plane 4. This scan is obtained by moving the transducer farther caudally (Fig. 4.49 a–c). A continuous midline echo is no longer demonstrated at this level, and the structures of the cerebellum appear prominently in the posterior fossa. Just anterior to the cerebellum are the bilobed structures of the cerebral peduncles, with a bright central echo representing the cerebral aqueduct. Lateral to this the structures of the hippocampus are visible at the medial wall of the occipital horn. The thalami are no longer seen, and a small portion of the frontal horns is still visible anteriorly.

Transverse plane 5. At this level we are for the first time able to discern the presence of three cranial fossae (Fig. 4.50 a–c). Because of the thickness of the frontal bone, we recognize no intracerebral structures in the anterior fossa. The echogenic sphenoid bone forms a clear demarcation between the anterior fossa and middle fossa. The structures of the cerebral peduncles appear more prominently between the cerebellum and sphenoid bone, and the cerebral aqueduct is more distinct. The cisterna magna is visible in the posterior fossa, directly adjacent to the cerebellum. Transverse plane 5 also includes the major blood vessels that supply the brain (Fig. 4.51). The sites of entry of the internal carotid artery (ACI) are visible on the flat portion of the sphenoid bone, and the middle cerebral artery (ACM) passes between the frontal and temporal lobes along the sphenoid ridge to the lateral sulcus, supplying most of the lateral cerebral surface. The thick, unpaired trunk of the basilar artery (AB) lies in the interpeduncular cistern between the posterior portions of the sphenoid bone and cerebral peduncles. The posterior cerebral artery (ACP) passes around the peduncles as the terminal branch of the basilar artery to supply the occipital lobe and two-thirds of the temporal lobe. The superior cerebellar artery (ACS) passes somewhat farther posteriorly and caudally around the midbrain to the cerebellar surface beneath the tentorium.

In real-time examinations, the arterial vessels can be seen to pulsate and are easily demonstrated. They can even be traced some distance peripherally by moving the transducer while observing the pulsations. Johnson et al. (1980b) stressed the importance of the intracerebral vascular anatomy as a guide to the identification of fetal brain structures by ultrasound. Kier (1974) published a pioneering work on the phylogenetic and ontogenetic development of the fetal cerebral arteries.

Demonstration and measurement of the fetal cerebellum is gaining increasing

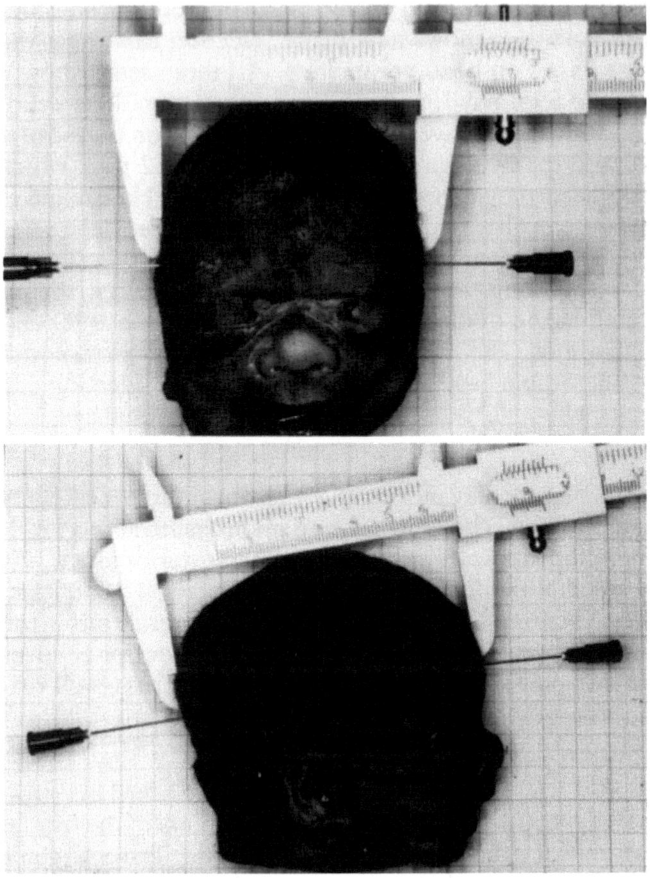

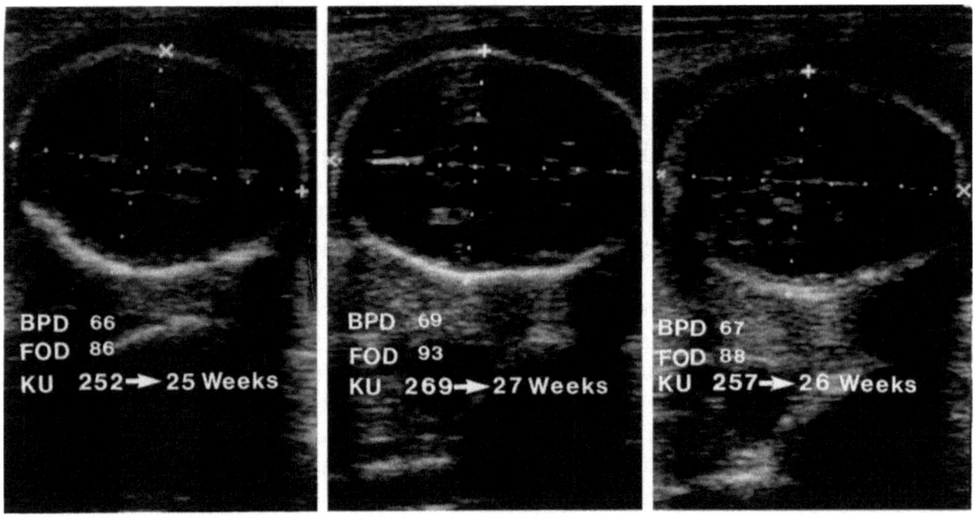
BPD 66
FOD 86
KU 252→25 Weeks
BPD 69
FOD 93
KU 269→27 Weeks
BPD 67
FOD 88
KU 257→26 Weeks

a b c

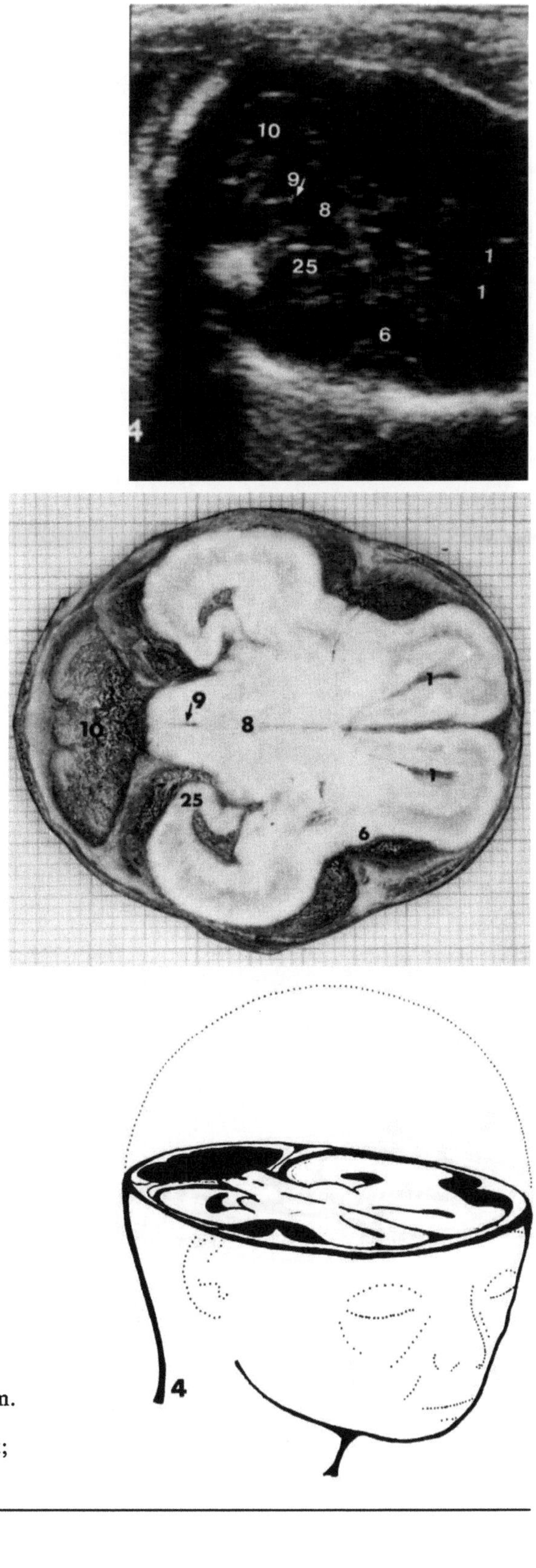

Fig. 4.49 a–c. Transverse plane 4.
a Sonogram. **b** Frozen section. **c** Diagram.
1 Frontal horn (lat. ventr.); *6* insula;
8 cerebral peduncle; *9* cerebral aqueduct;
10 cerebellum; *25* hippocampus

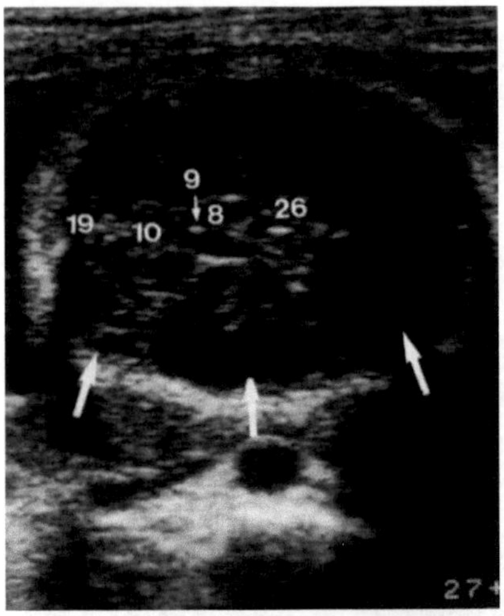

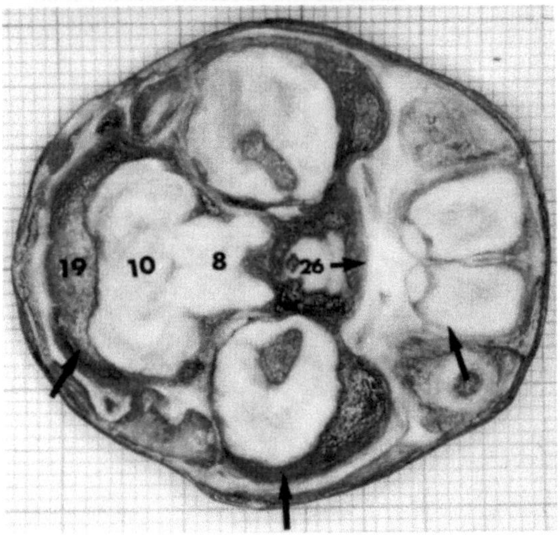

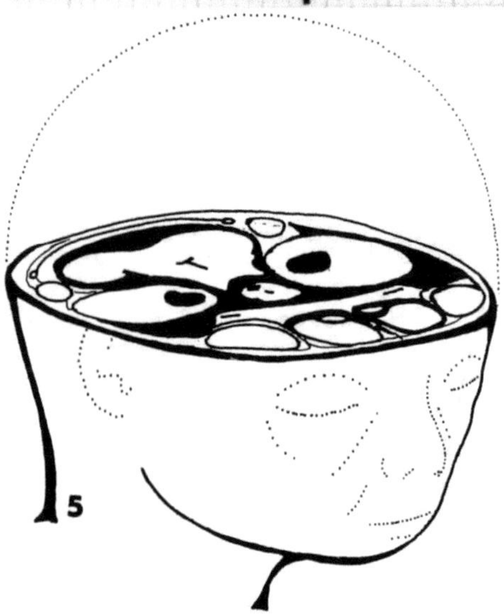

Fig. 4.50 a–c. Transverse plane 5.
a Sonogram. **b** Frozen section. **c** Diagram.
8 Cerebral peduncle; *9* cerebral aqueduct;
10 cerebellum; *19* cisterna magna;
26 sphenoid bone; ↑ cranial fossae

importance in prenatal ultrasound. The cerebellum is best demonstrated on a transverse scan performed through the acoustic window of the mastoid fontanelle (Figs. 4.52–4.54). Birnholz (1982), Campbell and Pearce (1983), and McLeary et al. (1984) have published on the measurement of the fetal cerebellum. McLeary et al. (1984) recommended measurement of the cerebellar diameter as an alternative to BPD measurement in cases of breech presentation, oligohydramnios, multiple gestation, and uterine anomalies. This recommendation is based on the findings of Hadlock et al. (1981b) and Kasby and Poll (1982), who showed that, with variations of the fetal head shape and compression, BPD as a biometric parameter is useless. The cerebellum, on the other hand, lies in the posterior fossa and is protected from compression by its proximity to the petrous ridges and occipital bone. Biometry of the cerebellum, then, is recommended in cases where the BPD appears unreliable.

In our study population, we were able to demonstrate and measure the fetal cerebellum in 95% of cases, confirming the observations of McLeary et al. (1984). The relationship between cerebellar diameter and gestational age was consistent with the results of Campbell and Pearce (1983) (Fig. 4.55). An increasing differentiation of cerebellar structures is apparent as gestation progresses. From 16 weeks on, the hemispheres and the vermis of the cerebellum are clearly seen (Fig. 4.52).

Accurate measurement of the cerebellar hemispheres relies on the proper selection of a clearly defined anatomic reference plane (transverse plane 5). At this level the cerebellum is bounded by the cisterna magna posteriorly, the petrous portion of the temporal bone laterally, and the cerebral peduncles anteriorly.

As gestation proceeds, the cerebellar structures become progressively more differentiated. From 24 weeks on we were able to demonstrate surface fissures with increasing frequency (Figs. 4.54 and 4.56). At about the same time, the cerebellar structures become differentiated into white matter and cortex (Fig. 4.57). Paramedian sagittal scans through the physiologic acoustic window in this region can also demonstrate the cerebellar folia (narrow cortical folds separated by fissures) (Fig. 4.58). The cisterna magna can be measured at the same level. Mahoney et al. (1984) described measurement of the cisterna magna in a series of 155 fetuses examined between 14 and 35 weeks. Midline measurements of the cisterna magna depth were made from the cerebellar vermis to the inner table of the occiput (Fig. 4.59). Statistical analysis demonstrated no significant change in the size of the cisterna magna between 15 and 36 weeks. The mean cisterna magna depth was 5 mm, with a range of 1–10 mm and one standard deviation of 3 mm. In 2 fetuses with Arnold-Chiari syndrome the cisterna magna measured only 2 mm or less in depth, althoug this did not allow antenatal diagnosis of the syndrome. The cisterna magna depth was consistently and markedly increased in fetuses with Dandy-Walker syndrome.

The fourth ventricle is demonstrated by rotating the scan plane downward at the back and upward at the front (Fig. 4.60), so that it approaches a more coronal orientation. In this plane the insula is again visible in the central part of the skull, and the frontal horns reappear anteriorly. The fourth ventricle is represented by a hypoechoic space with sharp posterior and lateral borders (Figs. 4.61 and 4.62). Only its transverse width can be measured, and this dimension showed no change between 19 and 36 weeks in a total of 211 cases. The mean width was 6 mm with a range of 3–7 mm (Fig. 4.62). In some cases we noted a circular structure located directly behind the cerebellar vermis (Fig. 4.63). This structure was described by Mahoney et al. (1984) and presumably represents the sinus rectus.

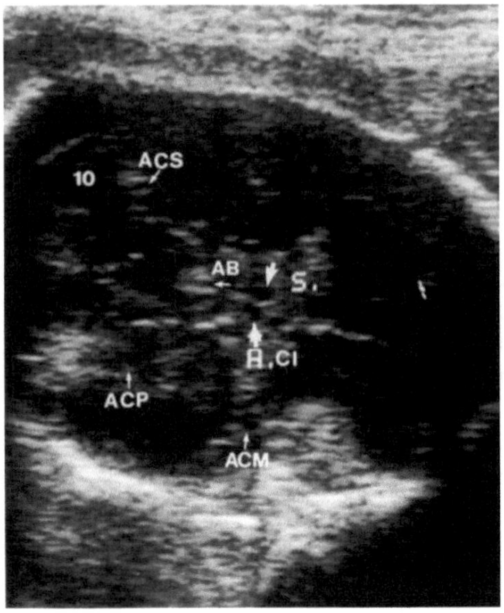

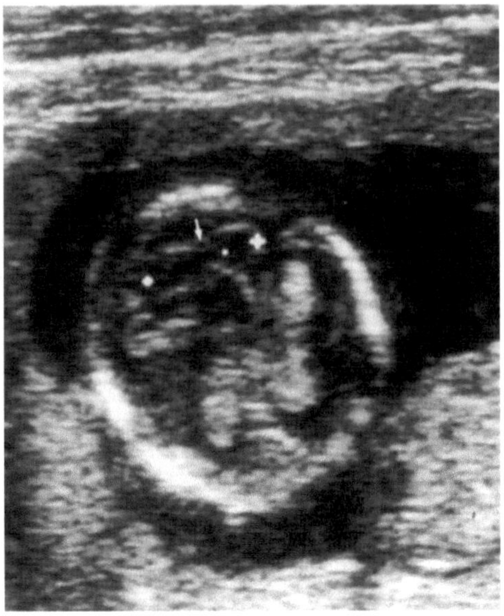

Fig. 4.51. (left) Location of the major brain arteries on transverse plane 5 (23 weeks). *S* Sphenoid; *AB* basilar artery; *ACI* internal cerebral artery; *ACM* middle cerebral artery; *ACP* posterior cerebral artery; *ACS* superior cerebellar artery

Fig. 4.52. (right) Appearance and measurement of the cerebellum at 15 weeks (transverse diameter 14 mm). The *arrow* marks the vermis of the cerebellum

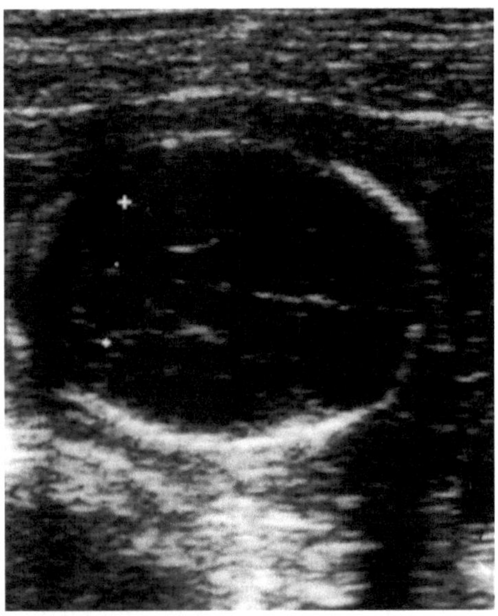

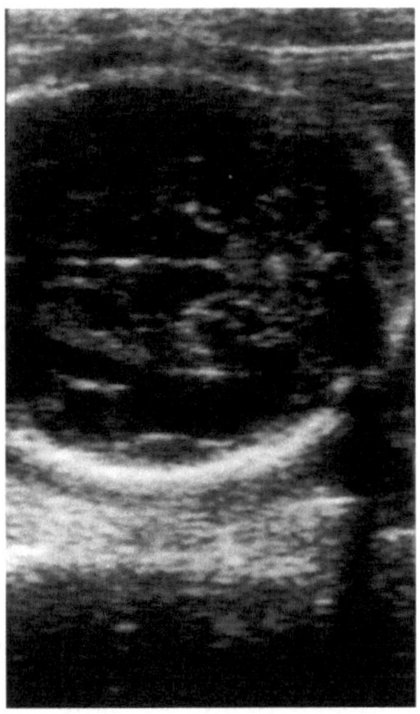

Fig. 4.53. (left) Measurement of cerebellar width at 17 weeks (18 mm)

Fig. 4.54. (right) Exceptionally good view of the cerebellar fissures obtained through the acoustic window of the mastoid fontanelle (23 weeks)

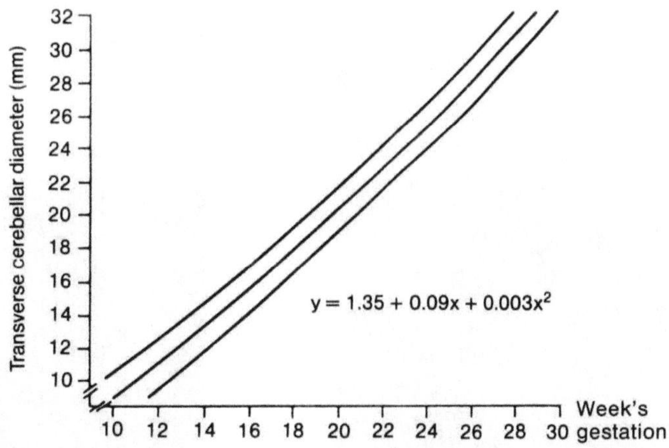

Fig. 4.55. Nomogram showing the relationship between cerebellar hemisphere width and gestational age (from Campbell) 1983)

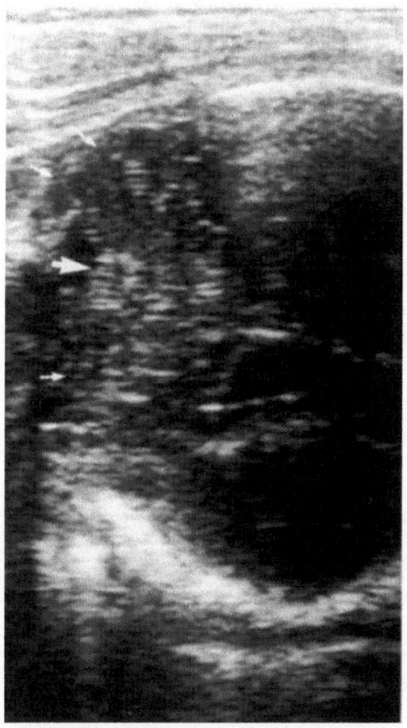

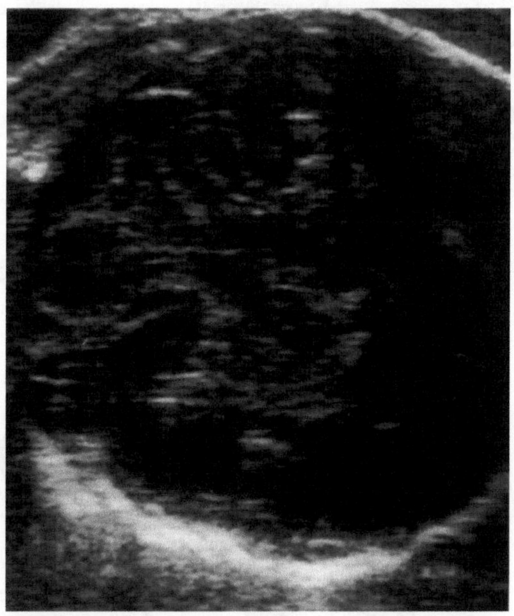

Fig. 4.56. (left) A scan of the cerebellum at 35 weeks shows a marked increase in the differentiation of the cerebellar surface. The *arrow* indicates the vermis

Fig. 4.57. (right) Scan of the cerebellum at 27 weeks, showing contrast between the cerebellar cortex and white matter

Transverse plane 6. This plane passes through the skull base, traversing the orbits anteriorly, the body of the sphenoid bone centrally, and the caudal portions of the cerebellum and cisterna magna posteriorly. The bony structures of the skull base that occupy this section prevent further differentation.

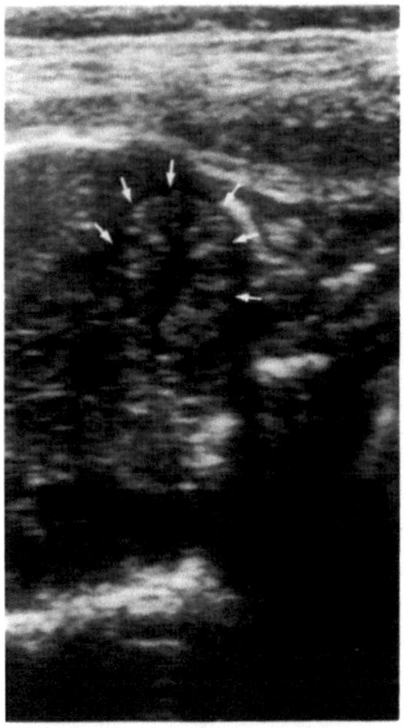

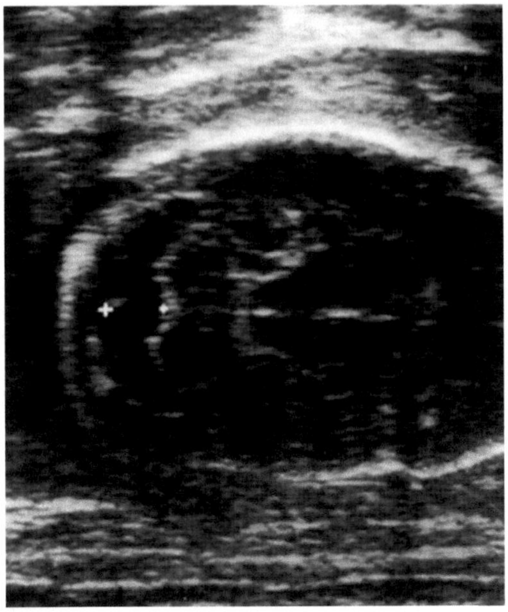

Fig. 4.58. (left) Paramedian sagittal scan through the cerebellum with an occipitoposterior head position (32 weeks). The *arrows* mark the foliae

Fig. 4.59. (right) Visualization and measurement of the cisterna magna on a transverse scan (20 weeks)

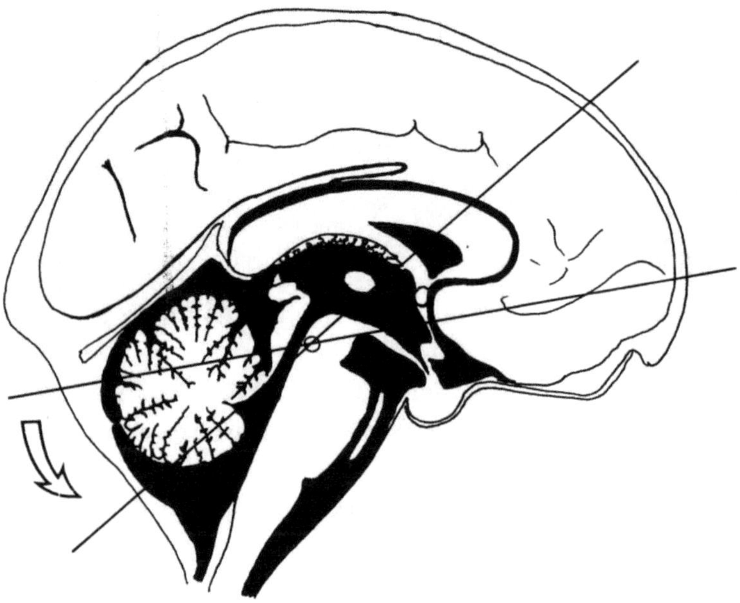

Fig. 4.60. Diagram showing how plane 5 is rotated to demonstrate the fourth ventricle

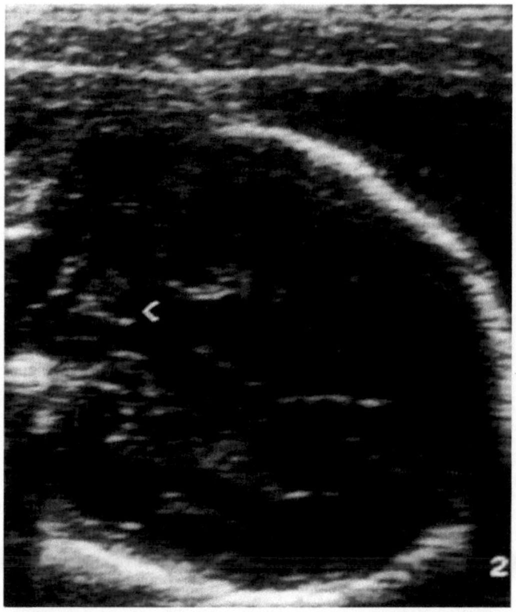

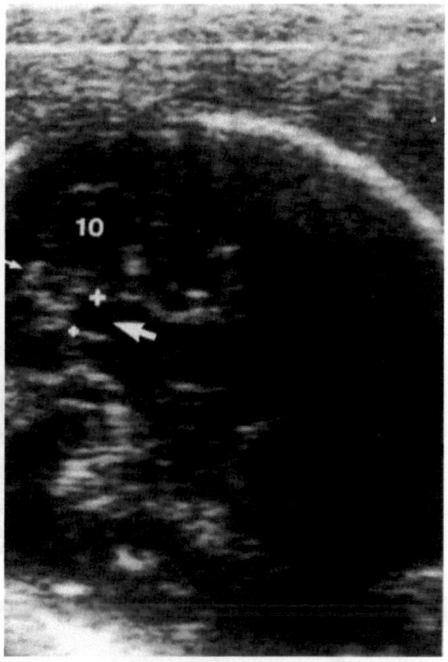

Fig. 4.61. (left) The fourth ventricle *(arrow)* is visible in this scan of the cerebellum

Fig. 4.62. (right) Measurement of the width of the fourth ventricle. The *small arrow* marks the cerebellar vermis

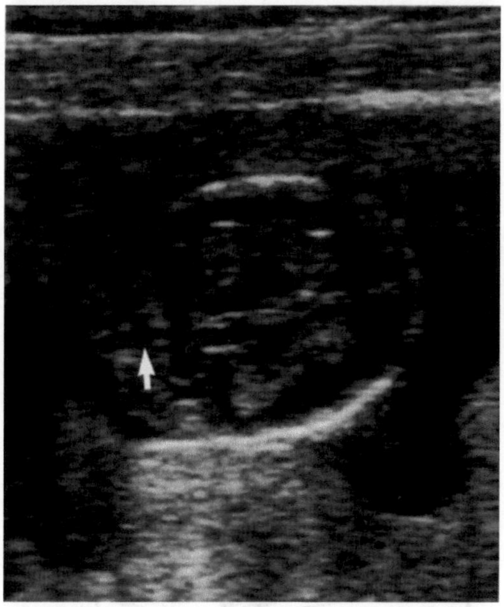

Fig. 4.63. Transverse scan through the sinus rectus *(arrow)* at 15 weeks

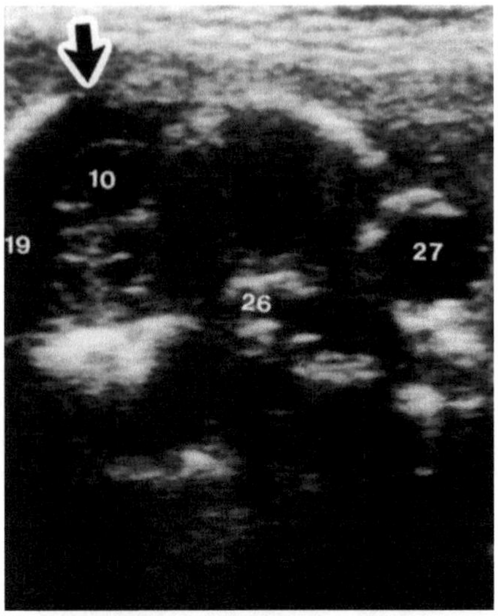

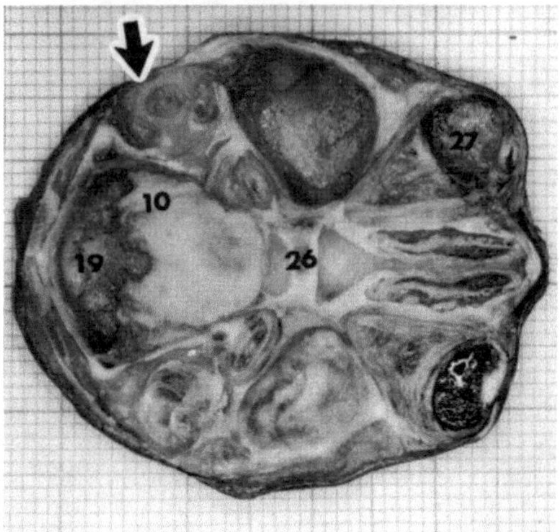

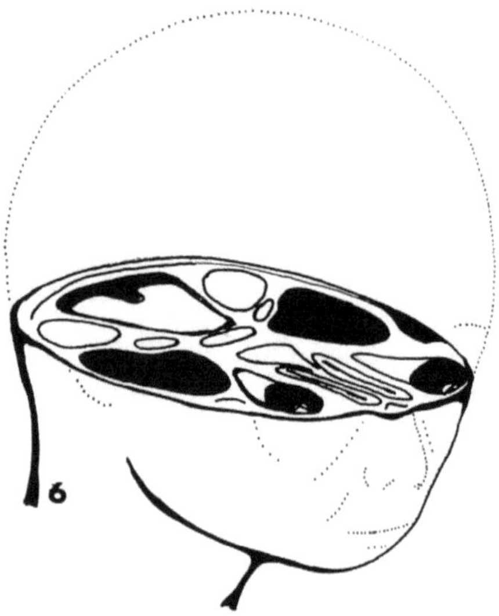

Fig. 4.64 a–c. Transverse plane 6.
a Sonogram. **b** Frozen section. **c** Diagram.
10 Cerebellum; *19* cisterna magna; *26* sphenoid bone; *27* orbit; ⟶ mastoid fontanelle

4.3.2 Sagittal Sections

Sagittal sections cannot be regarded as standard planes for intrauterine evaluation of the cerebrum. They depend strongly on the fetal presentation and are most readily obtained in breech presentations. Often the fetal head must be stabilized by palpation in order to obtain the desired scan plane. Once the sagittal plane has been established, however, the scans are of considerable value in providing detailed images of intracranial structures. Of particular diagnostic importance are the midsagittal scan (midline sagittal scan) and paramedian sagittal scans through the lateral ventricles.

Midsagittal section. Figure 4.65 a–c shows the midsagittal section on a sonogram, on a frozen section, and on an orientation drawing at 21 weeks. The sonographic landmark for confirming a midsagittal scan is the falx cerebri, which forms a crescent-shaped band with a uniform "ground-glass" appearance. The scalloped outlines of the cranial fossae are recognizable along the skull base. In the area of the facial skeleton, the scan must not traverse the orbits, as this would signify paramedian deviation of the scan anteriorly. The hypoechoic structure of the corpus callosum is visible anteriorly at the inferior border of the falx. The cavum septi pellucidi presents as a completely echo-free structure between the corpus callosum and fornix. Because the third ventricle is slit-like, and the beam focusing at this depth is no longer adequate to resolve such a narrow cavity, the structure of the thalamus generally appears below the fornix. The vascularized space of the cisterna venae magnae cerebri is seen posteriorly above the lamina tecti and the falx. In the posterior fossa, the cerebellum contrasts sharply with the less echogenic cisterna magna. At the skull base, the pons is visible behind the interpeduncular cistern, which contains the pulsating basilar artery. It is possible, then to identify all the principal midline brain structures on this section.

The midsagittal scan and associated key structures can be visualized from 15 weeks on (Fig. 4.66). In rare cases the inferior sagittal sinus and sinus rectus can be identified at the caudal border of the falx on a perfect midsagittal scan (Fig. 4.67 a, b). If the scan is moved even slightly from the midline, the texture of the falx disappears, the scan cuts the medial hemispheric surface tangentially, and gyri appear (Fig. 4.68). Birnholz (1986) states that the gyri are important in the assessment of fetal brain maturity. As gestation increases, a progressive differentation of the individual gyri can be observed in this plane (Figs. 4.69–4.70). Dorovini-Zis and Dolman (1977) documented the development of the cerebral gyri as a function of gestational age. They found that the calcarine and parieto-occipital sulci are already present at 21 weeks. We personally have not observed these features until 23 weeks (Fig. 4.68). The authors further report that the cingulate sulcus appears by 23 weeks, which is consistent with our findings. The cerebral surface undergoes a developmental "spurt" between 27 and 29 weeks, marked by the appearance of numerous new sulci and gyri. The changes are less striking between 29 and 39 weeks, with the first tertiary sulci appearing by 39 weeks (Figs. 4.69 and 4.70). The demonstration of these features may offer a new sonographic approach to the assessment of fetal brain maturity.

Paramedian sagittal section. This plane, illustrated in Fig. 4.71 a–c, can be found by moving the transducer laterally from the midsagittal plane and angling it posteriorly to obtain a complete image of the ventricular system. It should be possible to identify the frontal, occipital, and inferior horn, with the choroid plexus producing

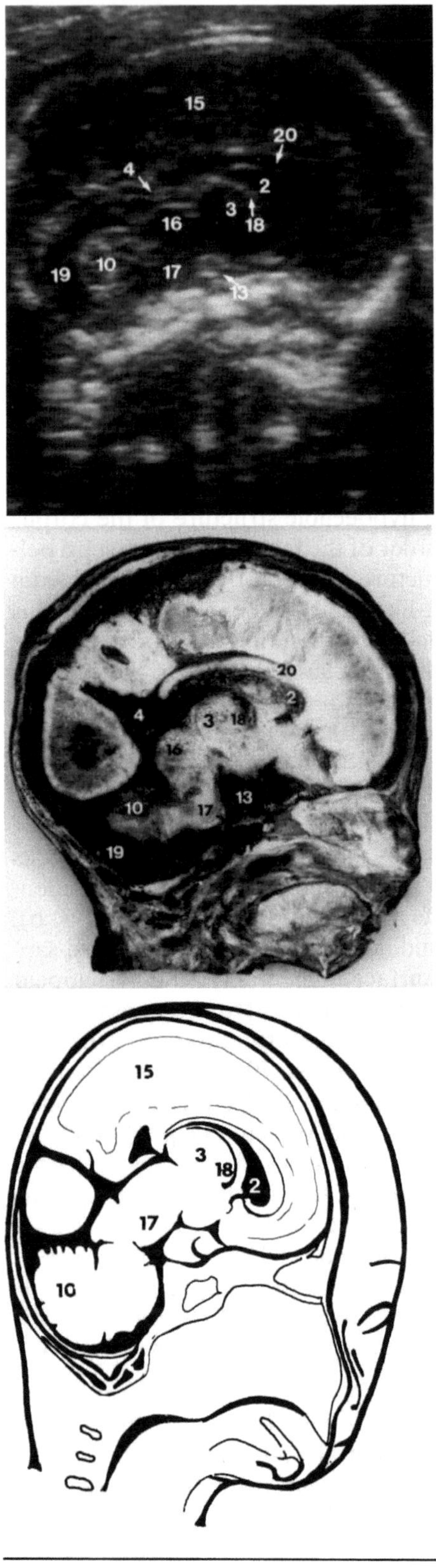

Fig. 4.65 a–c. Sagittal plane 1. Midsagittal section. **a** Sonogram. **b** Frozen section. **c** Diagram. *2* Cavum septi pellucidi; *3* thalamus; *4* cisterna venae magnae cerebri; *10* cerebellum; *13* interpeduncular cistern; *15* falx cerebri; *16* lamina tecti; *17* pons; *18* fornix; *19* cisterna magna; *20* corpus callosum

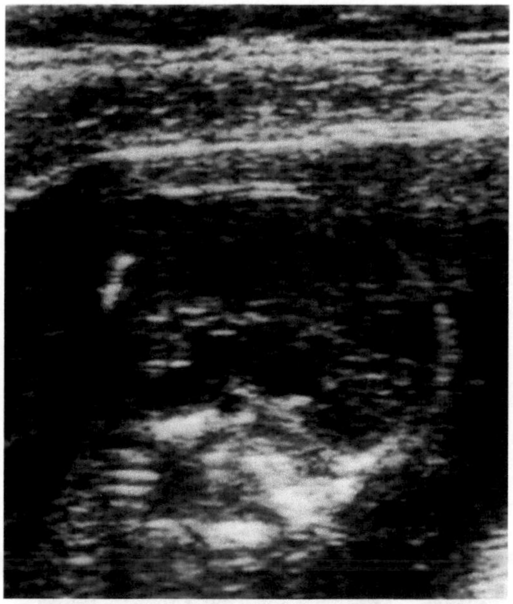

Fig. 4.66. Midsagittal scan at 16 weeks

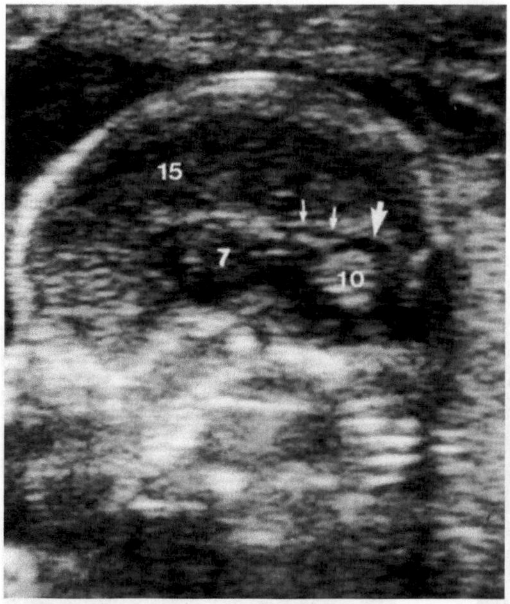

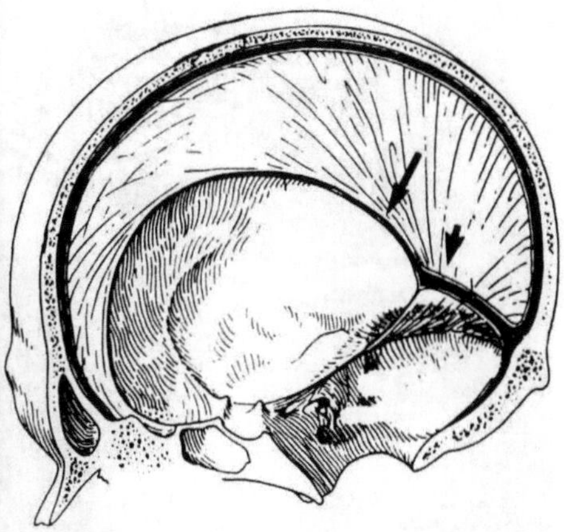

Fig. 4.67. a Midsagittal scan at 17 weeks (with the fetal head facing left). The *arrows* indicate the superior sagittal sinus and sinus rectus

Fig. 4.67. b Schematic drawing of the scan plane in Fig. 4.67 a. The *large arrow* points to the superior sagittal sinus, the *small arrow* to the sinus rectus

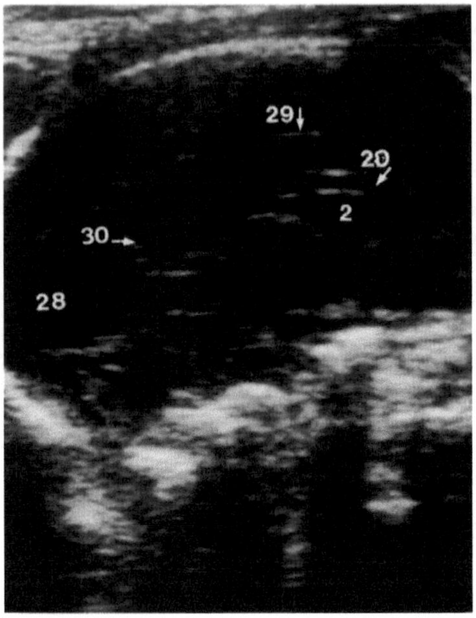

Fig. 4.68. Sagittal scan obtained slightly off the midline. The posterior part of the scan cuts the hemisphere tangentially; the gyri are demonstrated (23 weeks)

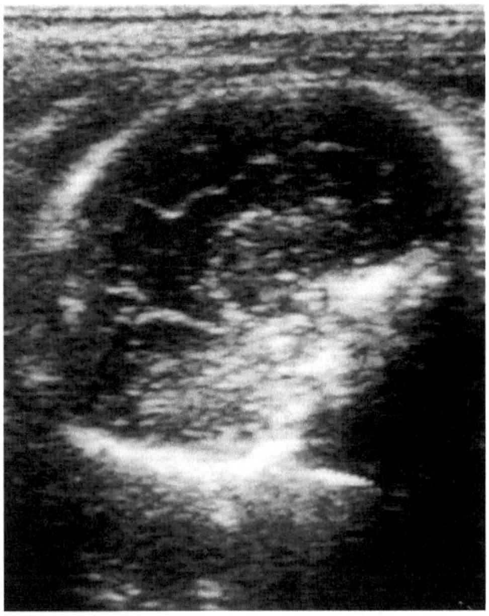

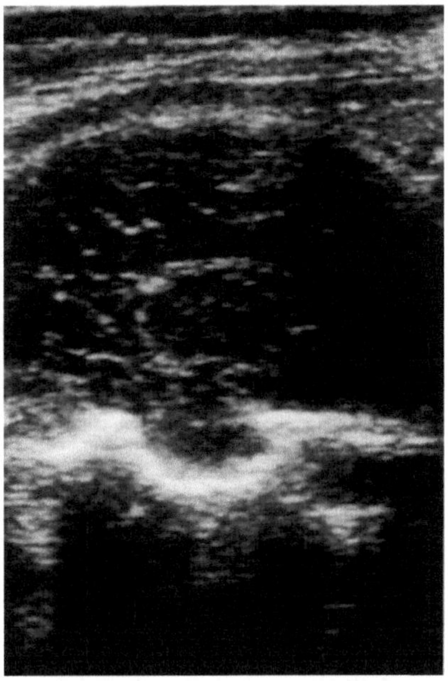

Fig. 4.69. (left) Tangential sagittal scan through a medial hemisphere at 32 weeks. The gyri are more numerous

Fig. 4.70. (right) Tangential sagittal scan through the medial part of the hemisphere at 39 weeks. Further differentiation of the gyri is apparent

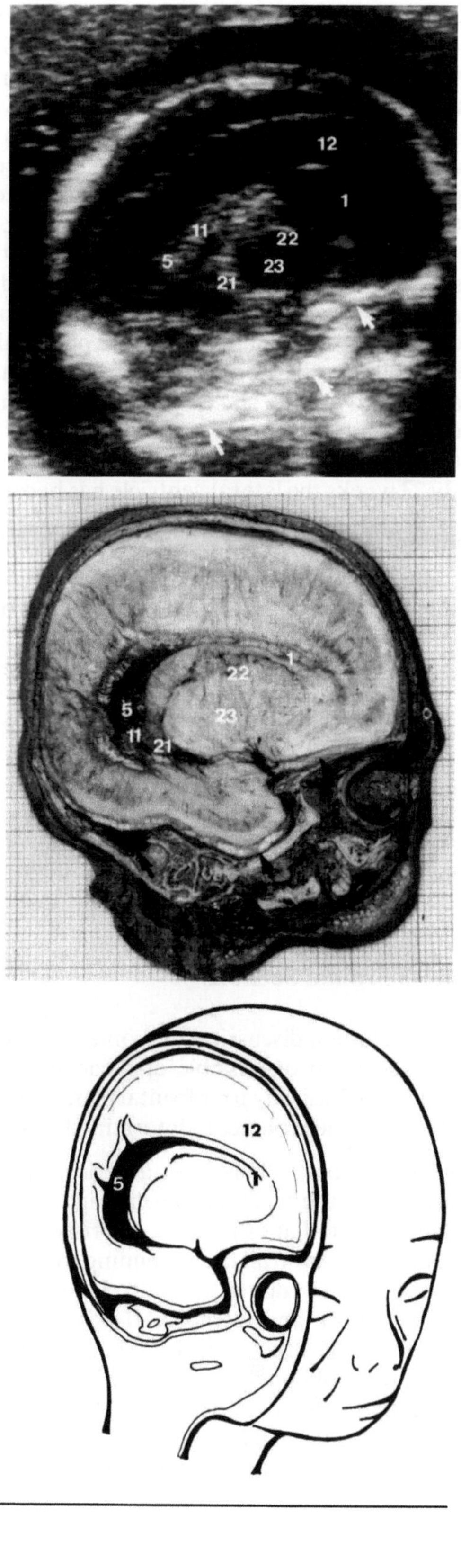

Fig. 4.71 a–c. Sagittal plane 2, paramedian sagittal section. **a** Sonogram. **b** Frozen section. **c** Diagram. *1* Frontal horn (lat. ventr.); *5* occipital horn (lat. ventr.); *11* choroid plexus; *12* brain mantle; *21* inferior horn (lat. ventr.); *22* capsula interna; *23* lentiform nucleus; ↖ cranial fossae

bright echoes in the inferior horn and part of the occipital horn. The brain mantle surrounding the ventricles appears relatively echo-free, and the margins of the ventricular system are sharply defined cranially and posteriorly. The scan identifies the orbits anteriorly as well as the outline of the three cranial fossae. The progressive development of the brain mantle and differentiation of the ventricular system are particularly well appreciated in this section. Figure 4.72 a, b compares a sonogram taken at 13 weeks with a frozen section. The cerebral mantle is extremely thin, and echogenic choroid plexus completely fills the central and occipital portions of the lateral ventricle. The occipital horn is still poorly developed at this stage. In their comparative anatomic studies of brain development, McFarland et al. (1969) were able to show that the occipital horn is a specialized structure that occurs only in primates. Thus, the occipital horn may be considered the phylogenically youngest part of the cerebral ventricular system and therefore it differentiates at a relatively late stage of fetal development. By 17 weeks the inferior wall of the occipital horn is arched and displaced upward by the enlarging cerebellum (Fig. 4.73), but the choroid plexus still almost completely fills the occipital horn. By 23 weeks the lateral ventricle has become slit-like in the area of the frontal horn, while the occipital horn has differentiated further and is no longer completely filled by choroid plexus (Fig. 4.74). The free portion of the occipital horn presents as a triangular, echo-free structure whose base measures on average 9 mm deep at this stage. Figure 4.75 a, b shows the results of radiographic studies of lateral ventricular development performed by Kier (1977) between 14 (Fig. 4.75 a) and 23 weeks (Fig. 4.75 b). Kier (1977) found that as gestation progressed, the ventricular width steadily diminished with the expansion of the surrounding brain. To emphasize the importance of this plane for the early detection of hydrocephalus, Fig. 4.76 shows a similar scan in a fetus with early hydrocephalus, which was diagnosed largely on the basis of the occipital horn that was too large for gestational age.
On the whole, sagittal scans provide a valuable adjunct to scans in other planes, especially in the differential diagnosis of structural brain pathology, although as mentioned earlier, they cannot be obtained in all cases.

4.3.3 Frontal Sections

We include a discussion of frontal sections and their typical features to complete the spectrum of possible approaches to the evaluation of fetal brain anatomy. There are basically four frontal planes of section that can be reproducibly obtained. Access to these planes is determined by the same factors as for the sagittal sections.

Frontal plane 1. This most anterior plane traverses the frontal lobes of the cerebrum and the facial skeleton (Fig. 4.77 a–c). The only intracerebral structures demonstrated are the falx cerebri and the hypoechoic frontal horns of the lateral ventricles. As on all the remaining frontal scans, the superior sagittal sinus is visible at the cranial edge of the falx. Visualization of the facial skeleton is limited. The dense supra- and infraorbital bony structures prevent further differentiation of more caudal structures of the facial skeleton. The frozen section (21 weeks) is cut somewhat obliquely, so that the left half is at a slightly more anterior level than the right half. The section cuts the globe within the right orbit, and in the left orbit we see the ocular muscles grouped around the optic nerve. The tongue also is visible on the frozen section inferior to the choanae.

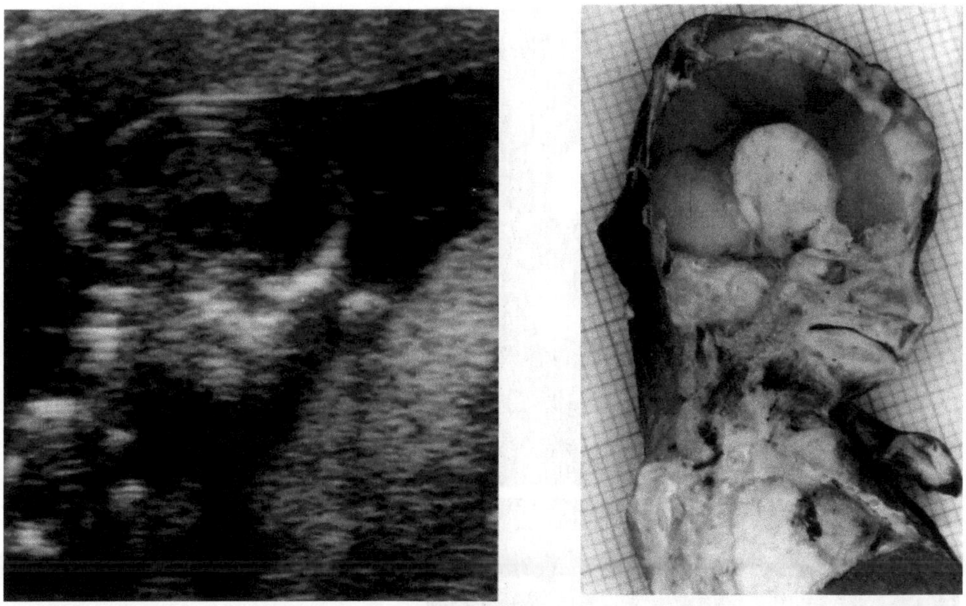

Fig. 4.72. a Paramedian sagittal scan through a fetal head at 13 weeks. Choroid plexus is visible in the ventricular system, which is still quite large

Fig. 4.72. b Analogous frozen section

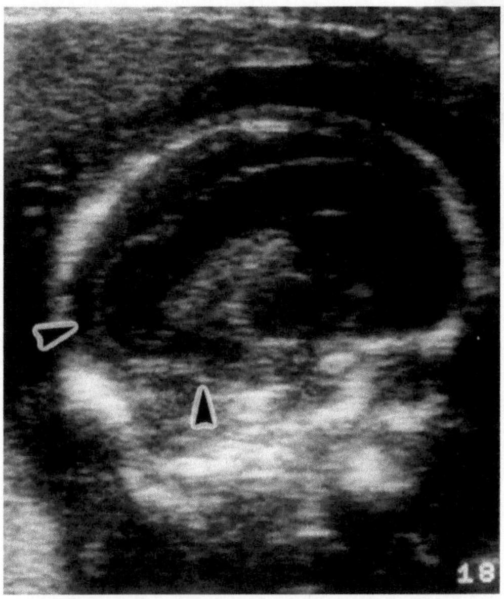

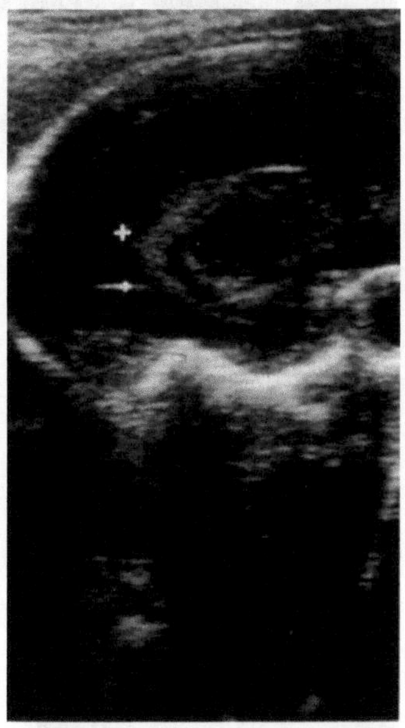

Fig. 4.73. (left) Paramedian sagittal scan through a fetal skull at 17 weeks. The *arrows* indicate the increasing differentiation of the occipital horn

Fig. 4.74. (right) Paramedian sagittal scan at 23 weeks. The points for measuring the depth of the occipital horn are shown

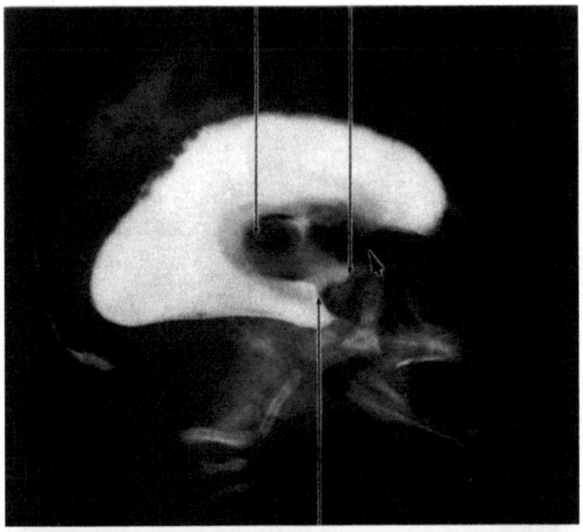

Fig. 4.75. a Lateral radiograph of a fetal head at 15 weeks showing the barium-filled lateral ventricles. The frontal horn is large, and the occipital horn is poorly differentiated at this stage (from Kier 1977)

Fig. 4.75. b Analogous radiograph at 23 weeks. The occipital horn shows marked differentiation compared to 15 weeks (Kier 1977)

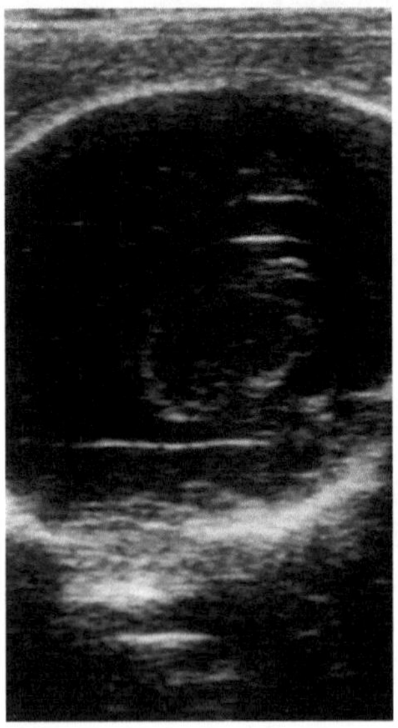

Fig. 4.76. Parasagittal scan through a brain at 33 weeks. The occipital horn is markedly distended signifying early hydrocephalus

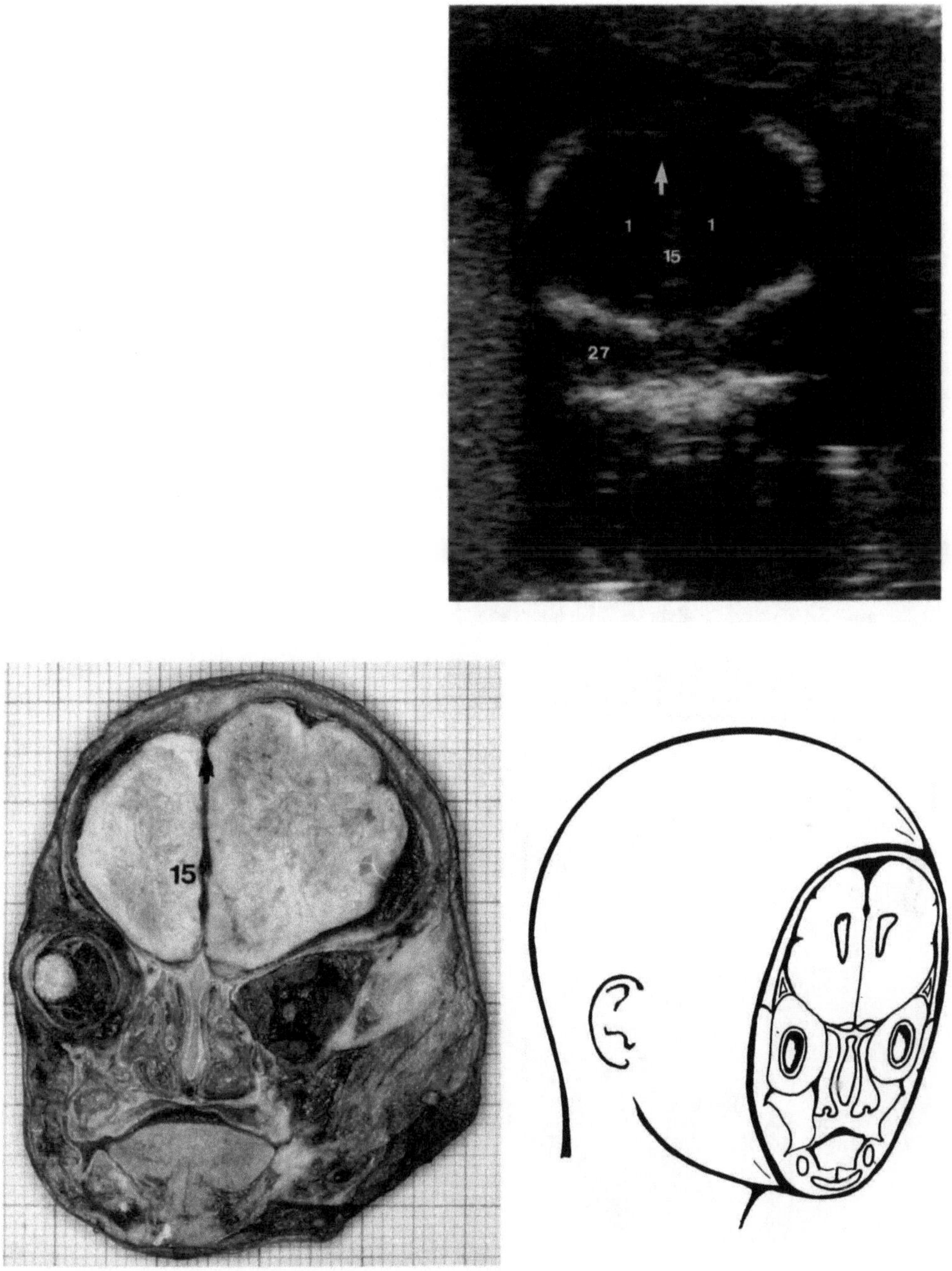

Fig. 4.77 a–c. Frontal plane 1. **a** Sonogram. **b** Frozen section. **c** Diagram. *1* Frontal horn (lat. ventr.); *15* falx cerebri; *27* orbit; ↑ superior sagittal sinus

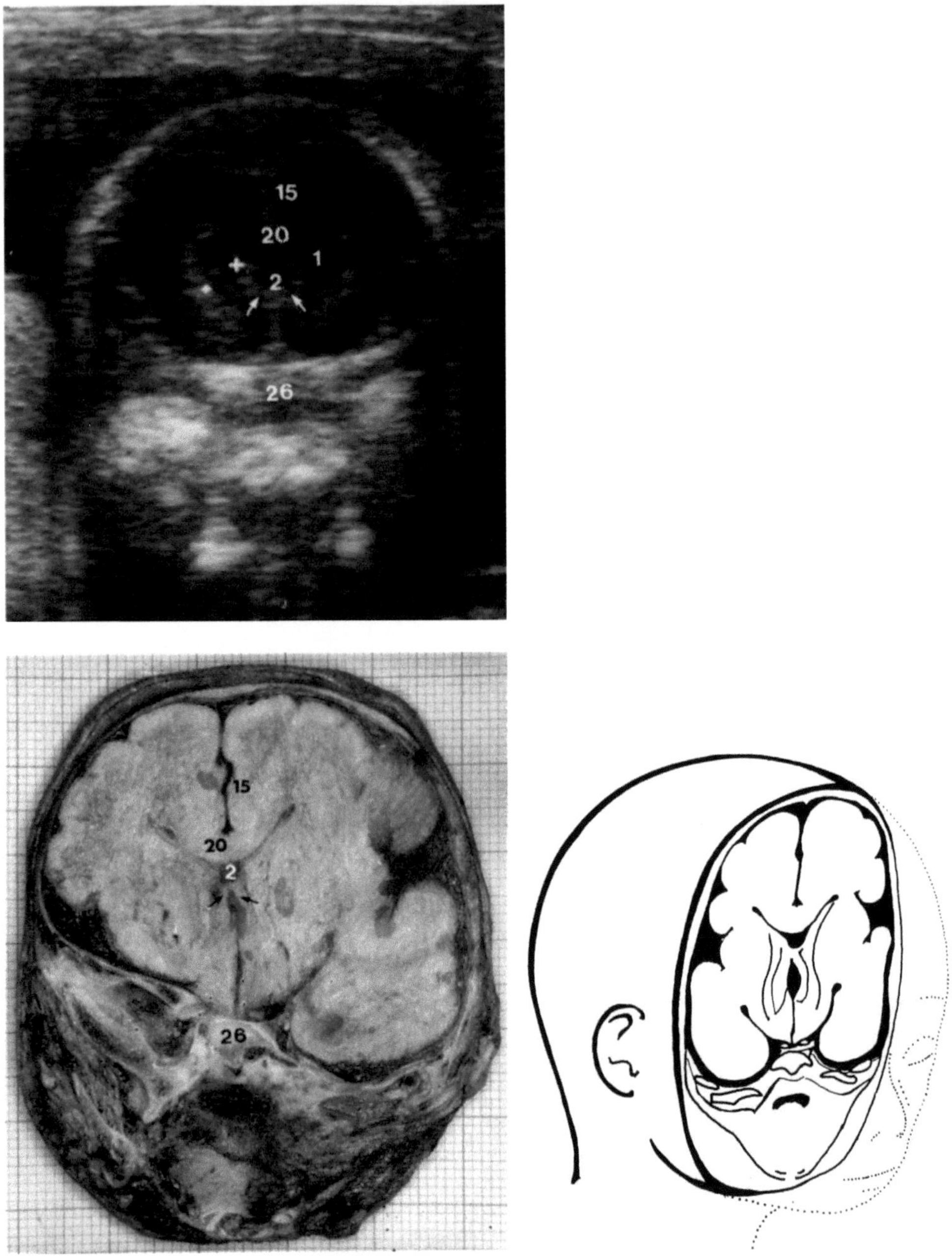

Fig. 4.78 a–c. Frontal plane 2. **a** Sonogram. **b** Frozen section. **c** Diagram. *1* Frontal horn (lat. ventr.); *2* cavum septi pellucidi; *15* falx cerebri; *20* corpus callosum; *26* sphenoid bone; ↖ fornices

Frontal plane 2 (Fig. 4.78 a–c). This plane is reached by moving the scan several millimeters posteriorly from level 1, leaving behind the basal structures of the frontal bone. The falx cerebri is seen dividing the cerebral hemispheres. The cavum septi pellucidi is visible between the frontal horns, and above is the corpus callosum. The fornices are visible in the frozen section, but on the sonogram they can only be assumed from anatomic relations. The dense structures at the skull base caudal to the lateral ventricles represent the sphenoid. It can be seen that the specimen was sectioned through the anterior fossa in its left half and through the middle fossa in its right half. The insula is not yet completely covered by brain structures at this stage. The lateral ventricles appear as slits on the frozen section, and they could not be dissected free in this plane in any of the specimens examined.

Frontal plane 3. This somewhat more posterior scan provides a detailed view of the structures of the midbrain (Fig. 4.79 a–c). It cuts the lateral ventricles at the junction between their central and occipital portions, and caudally we see the thalami and the cerebral crura, which are visible over their full longitudinal extent. Lateral to the crura is the inferior horn, bounded medially by the hypoechoic region of the hippocampus. McGahan et al. (1983) pointed to the importance of the fetal hippocampus and the possibility of its misinterpretation. The dense osseous structures of the skull base are represented by the portions of the petrous bone that appear on this section. The transducer can be rotated from this plane on a right-left axis through the cerebral peduncles to obtain an excellent view of the peduncles and cerebral aqueduct, with the interpeduncular cistern below and the cisterna venae magnae cerebri above (Fig. 4.80). Lateral to that is the anterior part of the cerebellar tentorium, which is bordered laterally by the occipital horns with their choroid plexus.

Frontal plane 4. A scan in this most posterior plane demonstrates the occipital lobes lateral to the cerebellar tentorium, with the occipital horns and choroid plexus at their center (Fig. 4.81 a–c). The cerebellum presents centrobasally and is flanked by the petrous ridges. The sinus rectus is visible at the free cranial border of the falx cerebelli. The progressive differentiation of the occipital horn also is apparent in this section, especially with regard to its size. Figure 4.82 a, b shows the occipital horns at 16 weeks as they appear sonographically and in frozen section. The occipital horns are large and filled with choroid plexus, and the brain mantle is relatively thin. By 27 weeks the occipital horns appear circular and devoid of choroid plexus, while the brain mantle has increased markedly in thickness (Fig. 4.83).

In most cases transverse sections of the fetal head are sufficient for evaluating intracranial features. However, if pathology is noted and the structures show abnormal size or anatomic relation, multiple scans are needed for a complete evaluation.

Tables 4.4 and 4.5 show the important anomalies involving the fetal skull and brain that can be detected by ultrasound. An important factor should be emphasized in connection with ultrasound evaluation of the central nervous system: While some congenital malformations based on defects in early morphogenesis usually result in severe distortion of brain anatomy and often are incompatible with life, other malformations are based on anomalies of differentiation and maturation caused by the action of environmental factors during later intrauterine growth and may produce only minor (if any) structural alterations. Such defects are often so subtle that they produce no recognizable morphologic changes on gross examination. Thus, a "normal" sonographic presentation by no means guarantees that the infant will have normal neurologic function.

Kurtzke et al. (1973) found in their studies that one-third of all congenital malformations recognized in the perinatal period involve the central nervous system. International reports on the incidence and mortality rate for all nervous system

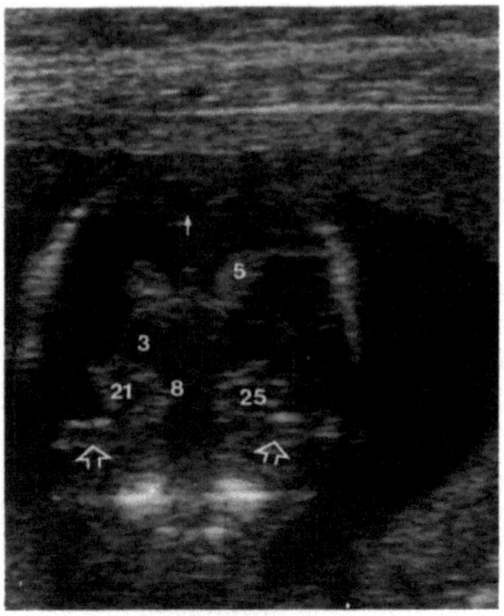

a

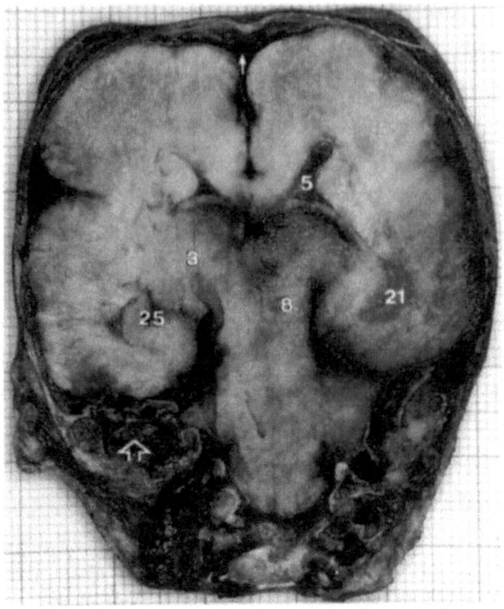

b

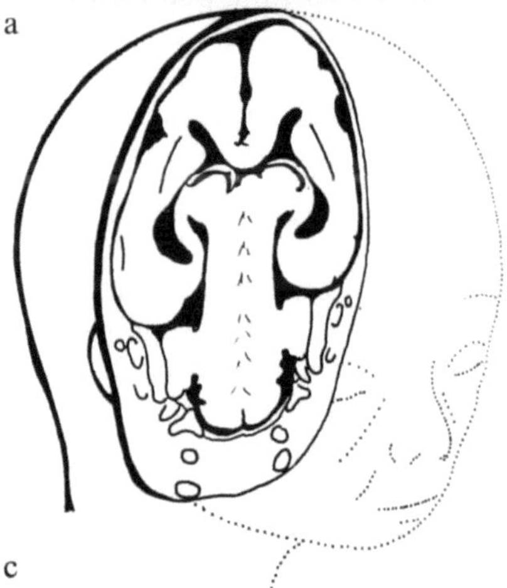

c

Fig. 4.79 a–c. Frontal plane 3. **a** Sonogram. **b** Frozen section. **c** Diagram. *3* Thalamus; *5* occipital horn (lat. ventr.); *8* cerebral peduncle; *21* inferior horn (lat. ventr.); *25* hippocampus; ↑ superior sagittal sinus; ⇧petrous part of temporal bone

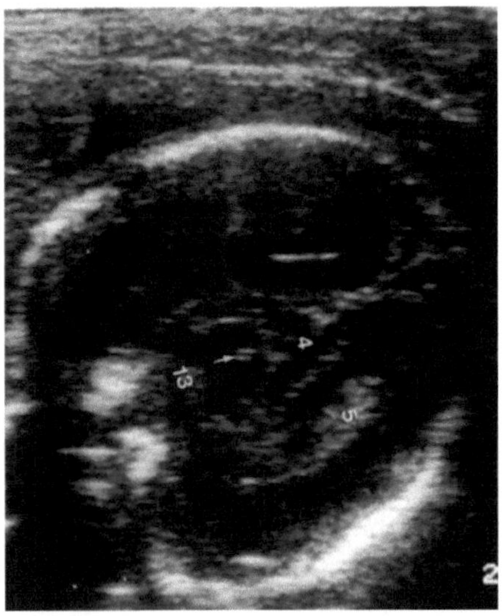

Fig. 4.80. Frontal scan at 21 weeks showing the cerebral peduncles and cerebral aqueduct

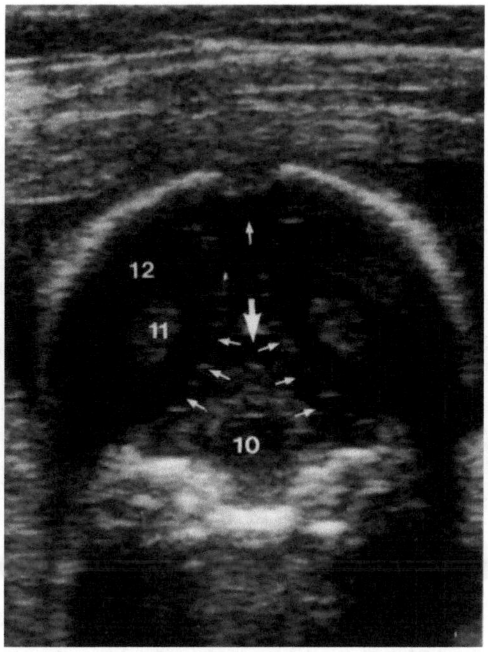

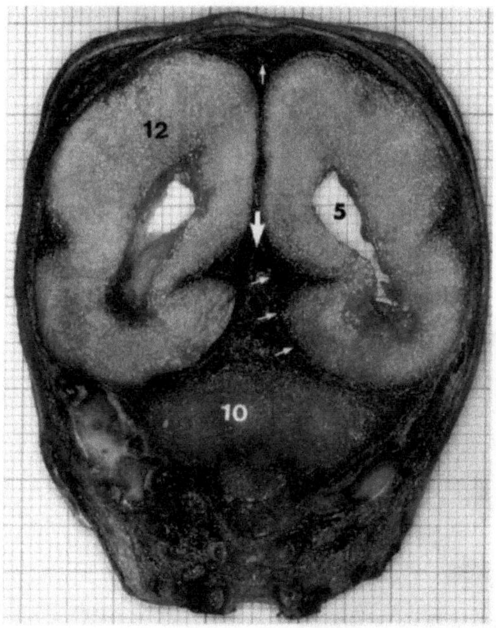

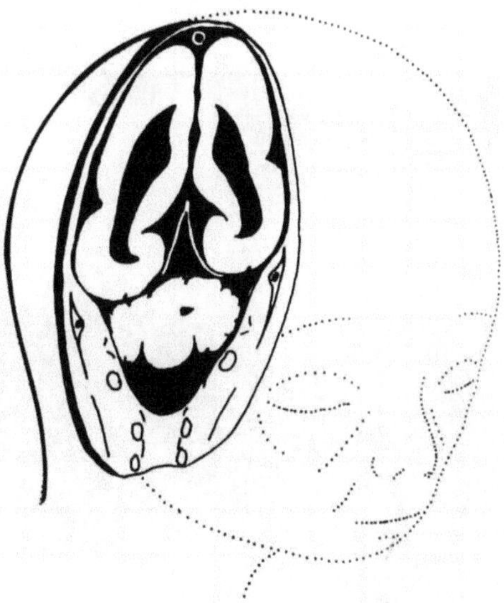

Fig. 4.81 a–c. Frontal plane 4. **a** Sonogram. **b** Frozen section. **c** Diagram. *5* Occipital horn (lat. ventr.); *10* cerebellum; *11* choroid plexus; *12* brain mantle; ➡ cerebellar tentorium

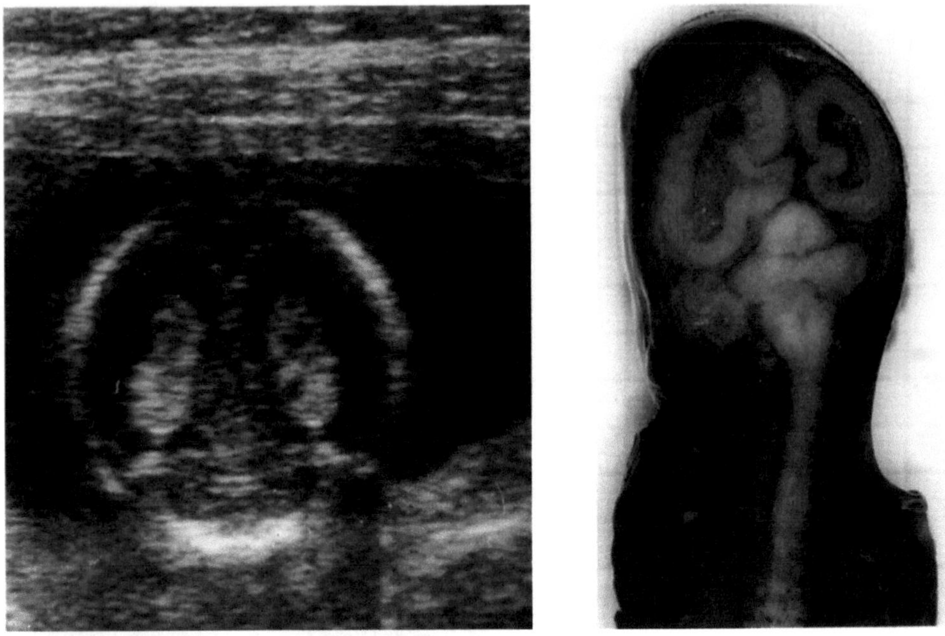

Fig. 4.82. a Frontal scan on plane 5 at 16 weeks. The ventricles are large and filled with choroid plexus, the brain mantle is relatively thin. **b** Analogous frozen section at 16 weeks

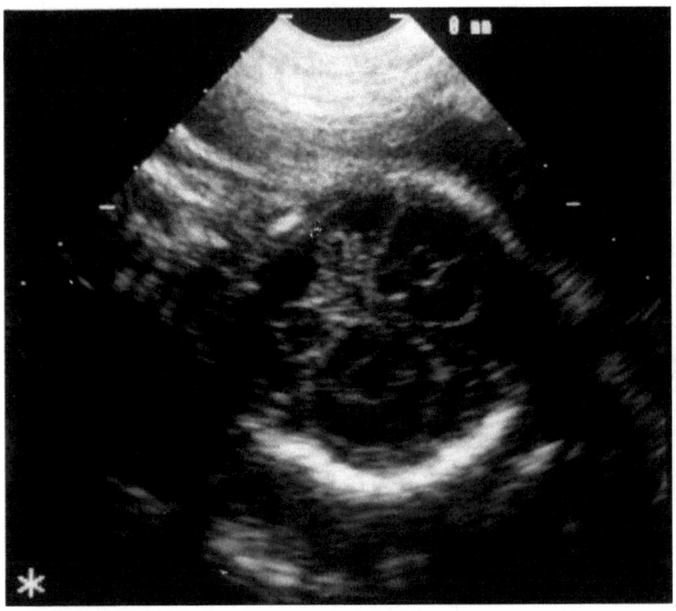

Fig. 4.83. Frontal scan on plane 5 at 27 weeks. The brain mantle has expanded. The cerebellum and cisterna magna are visible below the tentorium

Table 4.4. Classification of cerebral anomalies that can potentially be detected by ultrasound (modified from Hopf et al. 1984)

IV. Cerebral Anomalies

1. Failure of cranial closure
   a. Total cranioschisis
   b. Anencephaly
2. Midline defects (prosencephalies)
   a. Holoprosencephalies–alobular h., semilobular h., lobar h. (types A–F), olfactory lobe agenesis
   b. Absence of corpus callosum
   c. Malformations of septum pellucidum
3. Malformations of the gyri and cortex
   a. Agyria, pachygyria
   b. Heterotopia
   c. Microgyria
   d. Porencephaly
4. Abnormalities of brain size
   a. Microcephaly
   b. Macrocephaly
   c. Hemimacrocephaly

II. Cerebellar, Brain Stem Anomalies

1. Dysraphias and midline defects
   a. Vermian agenesis
   b. Dandy-Walker syndrome
   c. Arnold-Chiari malformation (types I–IV)
   d. Encephalocranial disproportion
2. Abnormalities of growth
   a. Complete cerebellar agenesis
   b. Cerebellar hypoplasia
3. Cerebellar cortical anomalies
4. Anomalies of the brain stem

III. Cerebral Anomalies of Genetic Etiology

1. Trisomy 21
2. Trisomy 18
3. Trisomy 13
4. Cri du chat syndrome
5. Klinefelter syndrome
6. Turner syndrome
7. XYY syndrome

IV. Other Syndromes

1. Pierre-Robin syndrome
2. Cornelia de Lange syndrome
3. Rubinstein-Taby syndrome
4. Brain malformations associated with craniofacial anomalies
5. Alcoholic embryopathy
6. Diabetic embryopathy

malformations range from 0.5 to 6–7 per 100,000 total population, with most deaths from severe anomalies occurring in the first year of life. Brain malformations are categorized according to morphologic criteria in Table 4.4 (Hopf et al. 1984). An etiologic classification of hydrocephalus is presented in Table 4.5.

Table 4.5. Etiology and classification of hydrocephalus (Hopf et al. 1984)

Obstructive or noncommunicating hydrocephalus		Communicating hydrocephalus	
1.	*Congenital causes*	*1.*	*Congenital causes*
1.1.	Aqueductal occlusion or stenosis	1.1.	Arnold-Chiari malformations
1.1.1.	Forking	1.2.	Encephaloceles
1.1.2.	Constriction	1.3.	Arachnitides
1.1.3.	Transverse septum	1.4.	Lissencephaly
1.1.4.	Gliosis	1.5.	Congenital absence or dysplasia of Pacchioni granulations
1.2.	Atresia of the foramen of Luschka and foramen of Magendie: Dandy-Walker syndrome	2.	Acquired causes
1.3.	Space-occupying lesions	2.1.	Arachnitides
1.3.1.	Benign intracranial cysts	2.1.1.	Infection
1.3.2.	Tumors	2.1.2.	Hemorrhage
1.3.3.	Vascular malformations	2.1.3.	Particles (cellular) in cerebrospinal fluid
2.	*Acquired causes*	2.2.	Space-occupying lesions
2.1.	Aqueductal occlusion or stenosis (gliosis)	2.2.1.	Tumors
		2.2.2.	Nontumorous lesions
2.2.	Ventriculitis	2.2.3.	Platybasia
2.3.	Space-occupying lesions		
2.3.1.	Tumors	*3.*	*Overproduction of CSF*
2.3.2.	Nontumorous lesions	3.1.	Papillomas of the choroid plexus

4.4 Face

Hansmann et al. (1985) have emphasized the importance of the ultrasound evaluation of fetal facial structures. Detailed examination of the face is gaining increasing importance not only from the standpoint of morphologic assessment but also for recognizing movement patterns that aid in the evaluation and differentiation of physiologic states. Studies of fetal eye movements and swallowing movements are of particular interest in this regard (Bots et al. 1981; Birnholz 1981; Nihuis et al. 1982; Birnholz 1983). Some highly experienced examiners even report observing facial expressions, which apparently can be related to fetal condition (Hansmann et al. 1985).

Even in routine Stage 1 examinations, we recommend that attention be directed to facial detail because of their potential pathognomonic importance. Thus, facial clefts, micrognathia, hypo- and hypertelorism, and anomalies of the ears are commonly associated with chromosomal disorders and malformation syndromes, and it is not unusual for facial abnormalities to prompt a detailed search for more serious anomalies.

4.4.1 Sagittal Profile Scan

As mentioned earlier in this chapter, an attempt should be made to delineate the fetal profile before proceeding with detailed screening of the intracranial anatomy. This is done by positioning the transducer on the maternal abdomen such that the fetal head position is occipitoposterior relative to the transducer (Fig. 4.84). As most modern instruments have a zoom capability, the face can be selectively magnified and scrutinized (Fig. 4.85). It is imperative that an exact midsagittal scan be obtained. Any other scan may yield a bizarre and misleading "pseudoprofile" that may appear very abnormal. None of the three "profiles" in Fig. 4.86 a–c appears normal, because the scans were angled off the midline, as indicated by the presence of the orbit on all three scans. Obtaining the fetal profile from an oblique scan through the lateral part of the maxilla, for example, would create the impression of micrognathia in a two-dimensional projection (Fig. 4.87). A correct sagittal scan in the midline should demonstrate the frontal bone, nasal bone, maxilla, and mandible in their proper sequence.

An exact midsagittal scan can be confirmed by moving the transducer to the left and right and noting whether it moves equal distances from the midline before reaching the left and right orbits.

A fetal mouth-opening sequence is recorded in Fig. 4.88 a–c. Yawning movements of this type are commonly observed and tend to occur in episodes. The orbit is not seen on any of the scans, and all three scans depict the nasal bone and nasal soft tissues, the upper lip, and the lower lip. In Fig. 4.88 b and c the tongue is clearly visible within the open mouth.

4.4.2 Frontal View

A frontal view of the face is obtained by rotating the transducer 90° about the long axis of the skull and then angling it as necessary on the right-left axis to define facial soft-tissue structures in coronal section (Fig. 4.89). The scan plane may be progressively angled to obtain isolated views of the forehead and nose (Fig. 4.90 a, b) and of the nostrils, alae nasi, and lips with the mouth closed (Fig. 4.91 a, b) or open (Fig. 4.92 a, b). As the scan is angled farther caudally, cranial portions of the facial soft tissues disappear from the image as more caudal portions come into view. In Fig. 4.93 a, b, for example, only the tip of the nose is still visible when the chin area is reached. An isolated view of the lips is obtained only when the mouth is open (Fig. 4.94 a, b).

4.4.3 Orbits

A complete screening of the facial skeleton should include an evaluation of the bony structures of the orbit, maxilla, and mandible. This is not essential for basic screening, but it is absolutely necessary in patients at risk for fetal anomalies, especially when there is a question of faciocranial anomalies. It is best to demonstrate the orbits first before directing attention to the maxillae. For the less experienced examiner, we recommend starting at the level of the BPD (transverse plane 3) and proceeding as follows (Fig. 4.95):

In phase 1, the transducer is moved caudally to the level of transverse plane 6, which should demonstrate the three cranial fossae and the orbits anteriorly. In phase 2 the scan is angled upward posteriorly, using the orbits as a pivot point, until the cerebellum disappears from the image (Fig. 4.95 and Figs. 4.96 a–c). Mayden et al. (1982) described this reference plane as an acceptable plane for the measurement of orbital parameters. Jeanty (1984a) likewise determines the biorbital width

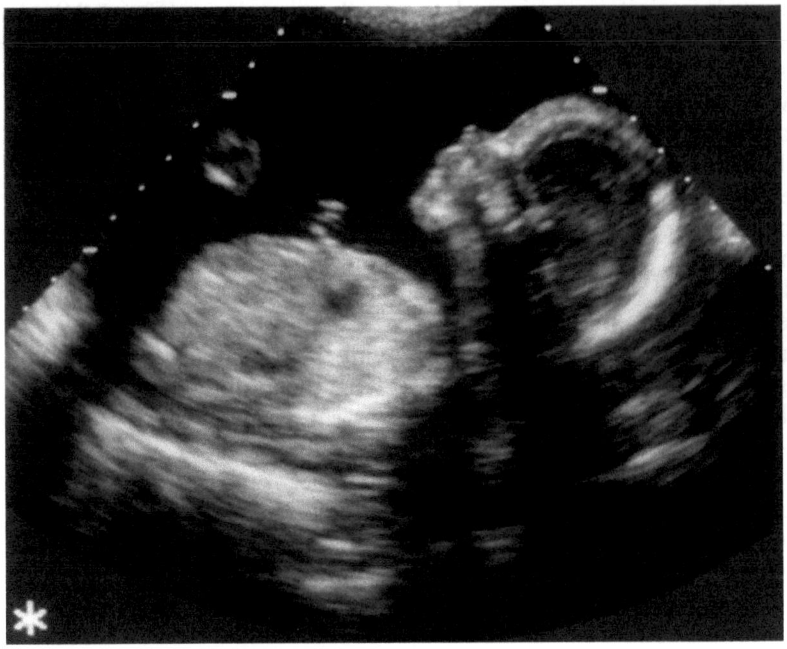

Fig. 4.84. Midsagittal scan through a 19-week fetus in the occiput posterior position. The profile is well defined

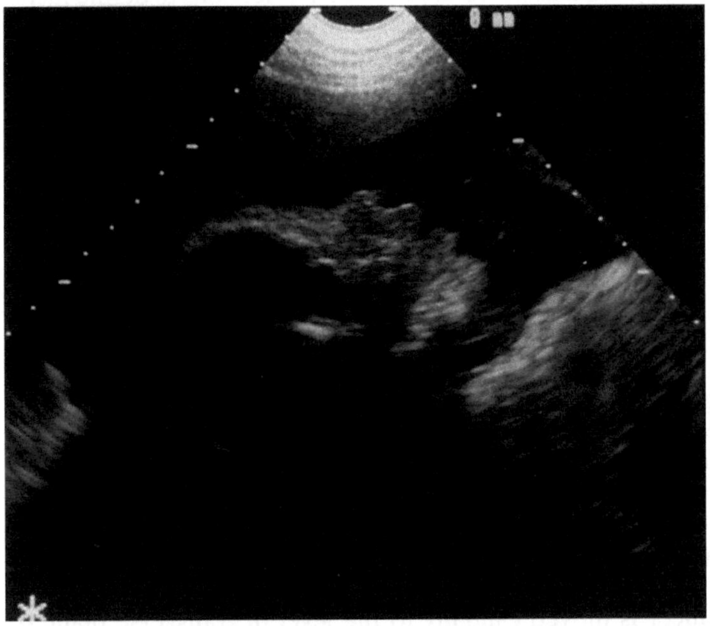

Fig. 4.85. Close-up view of the fetal profile. The mouth is partially open, the orbit is not visualized

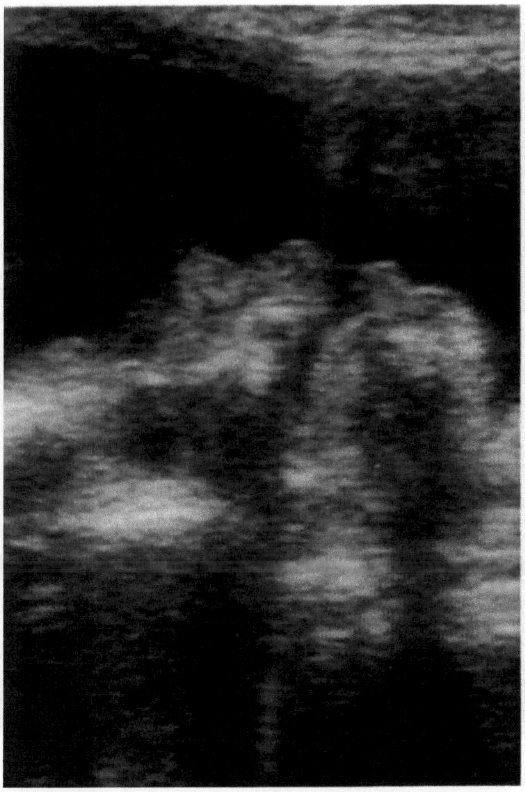
a

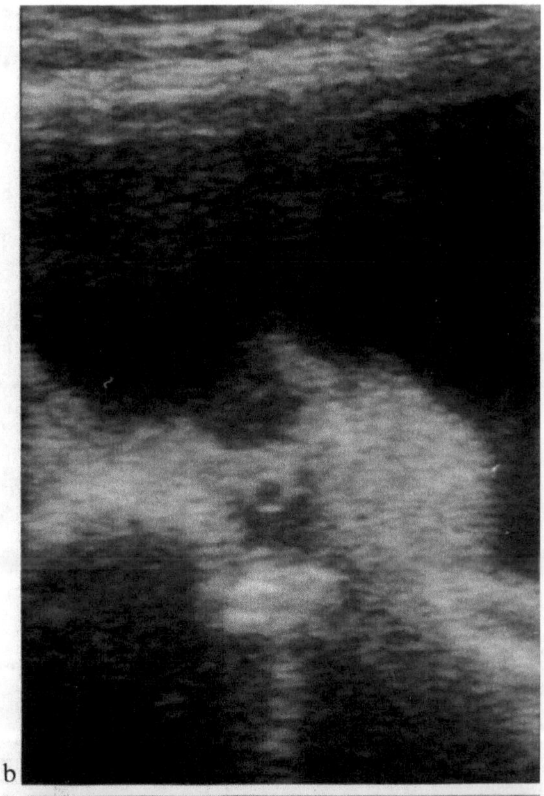
b

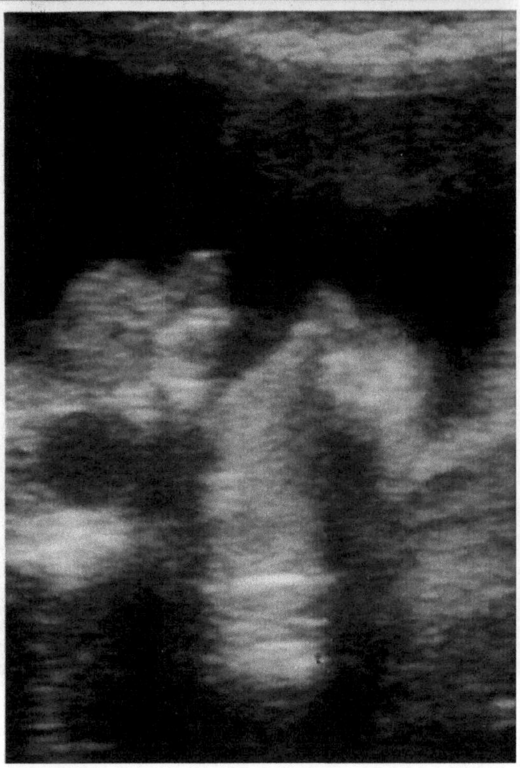
c

Fig. 4.86 a–c. "Pseudoprofile" views. The orbit is visible on all scans, none of which demonstrate the nasal bone. The scans are not midsagittal, giving the facial contour an abnormal appearance

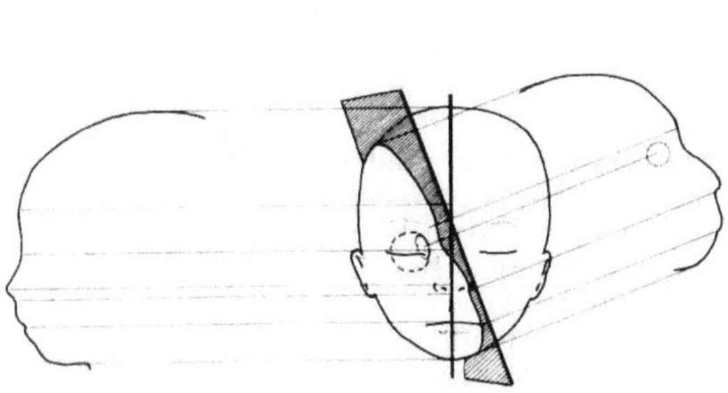

Fig. 4.87. Reconstruction of facial profiles from different scan planes. The profile from the midsagittal scan *(left)* appears normal. An oblique scan through the orbit *(right)* cuts the mouth tangentially, creating the impression of micrognathia

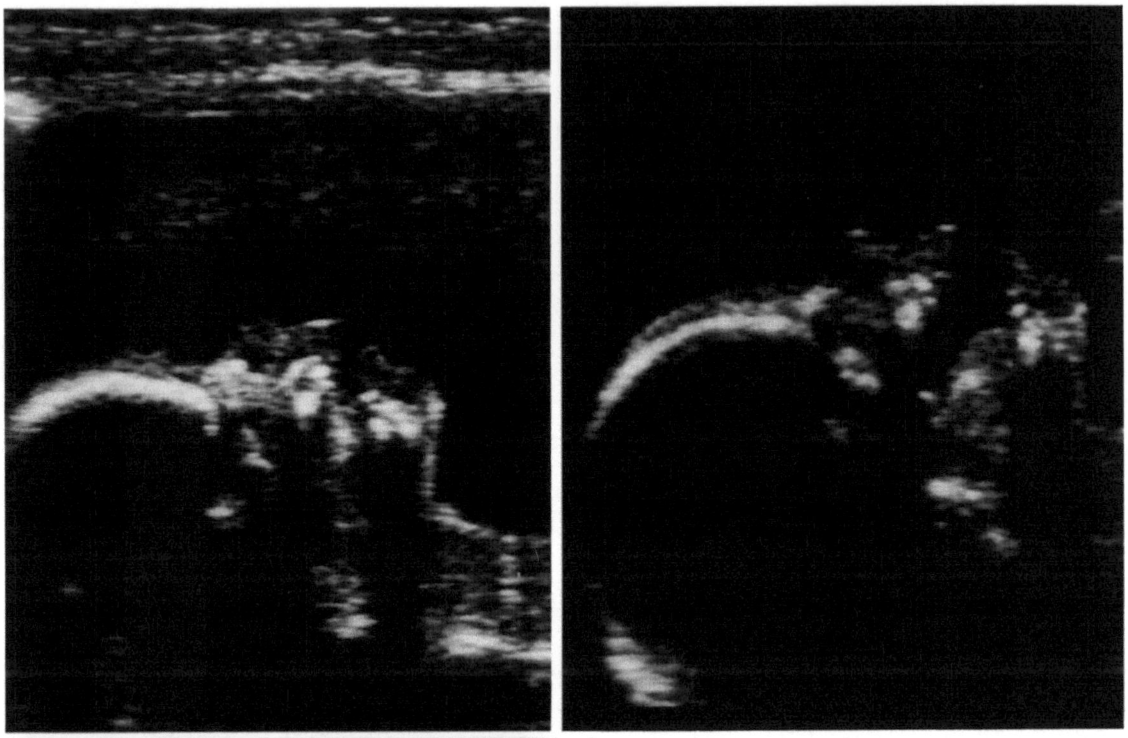

a b

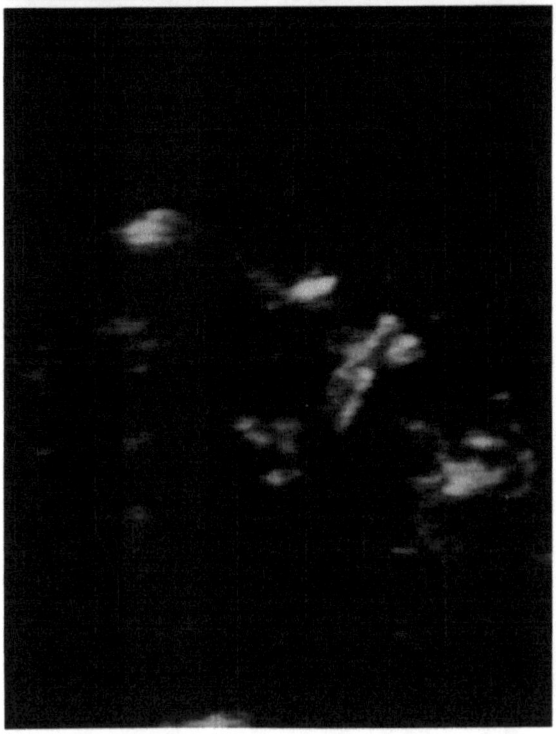

c

Fig. 4.88 a–c. Exact midsagittal profile scans. All the scans demonstrate the nasal bone; the orbit is not visible. **a** Mouth closed, maxilla and mandible visible behind upper and lower lips. **b** Mouth slightly open, tongue retracted, lip contours clearly defined. **c** Mouth wide open ("fetal yawn")

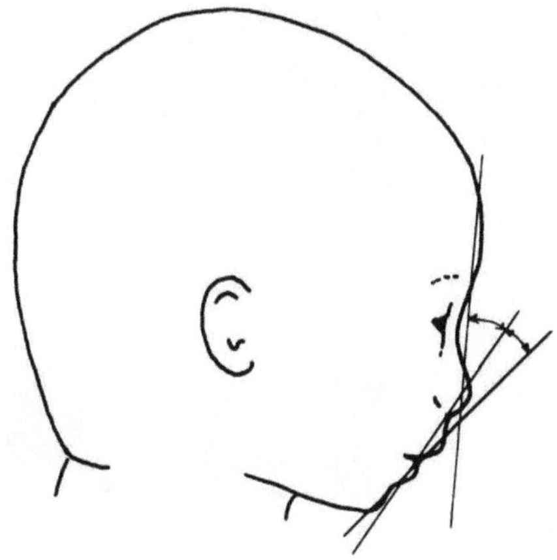

Fig. 4.89. Diagram showing frontal scans for examining the face. The scan plane is angled to obtain different sections of the face

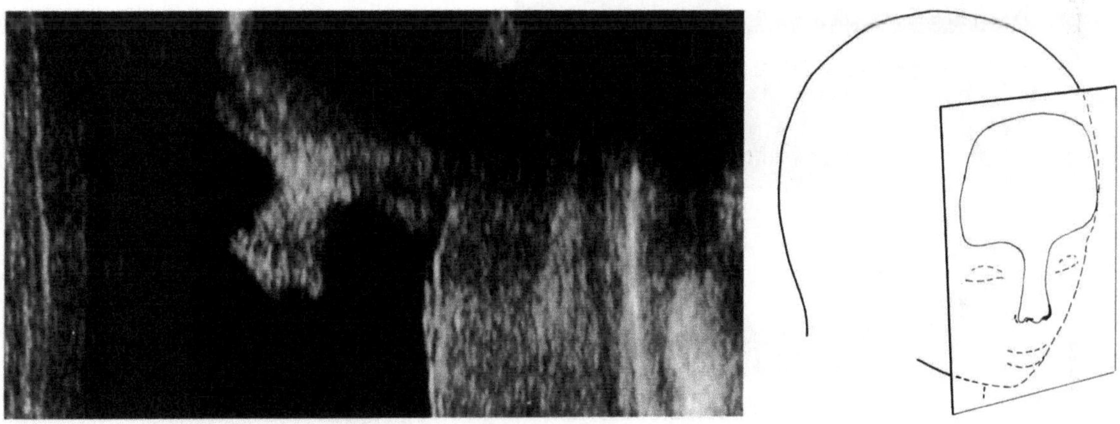

Fig. 4.90 a, b. Tangential frontal scan through the forehead and nose

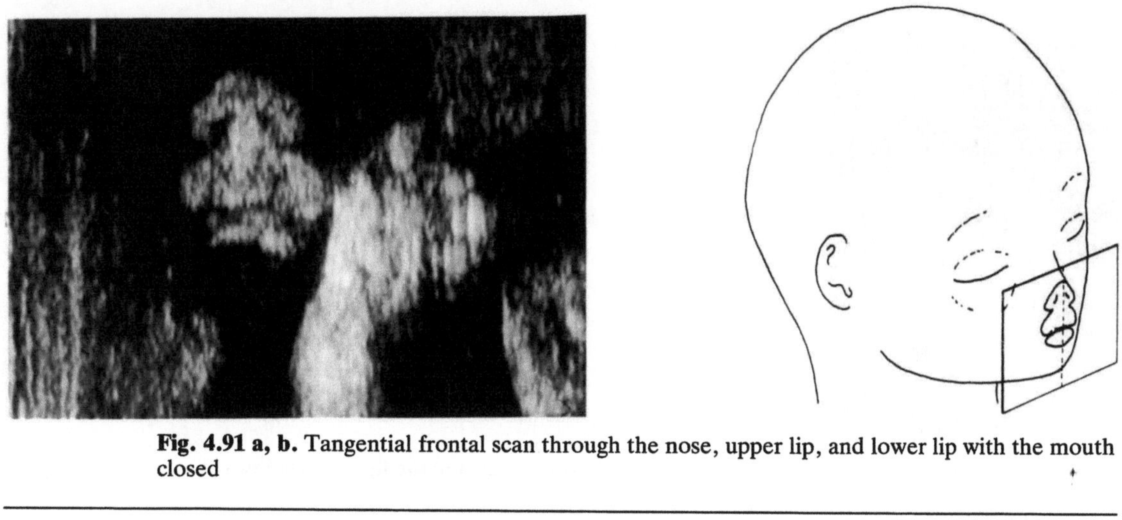

Fig. 4.91 a, b. Tangential frontal scan through the nose, upper lip, and lower lip with the mouth closed

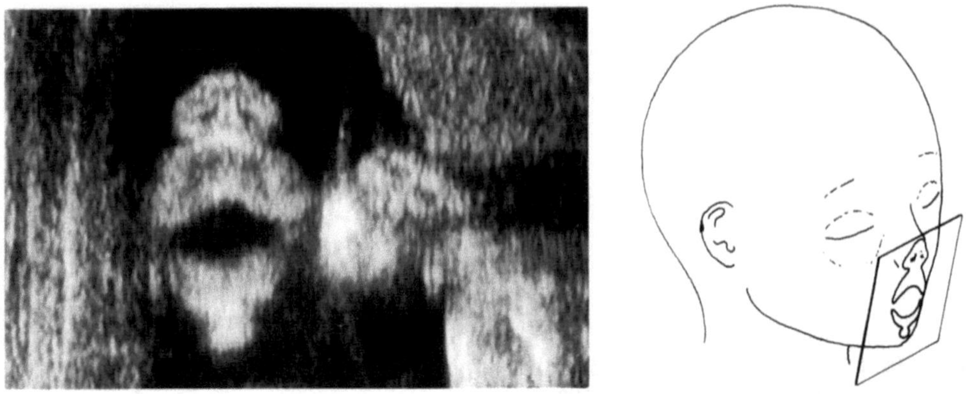

Fig. 4.92 a, b. Tangential frontal scan through the nose and lips with the mouth open

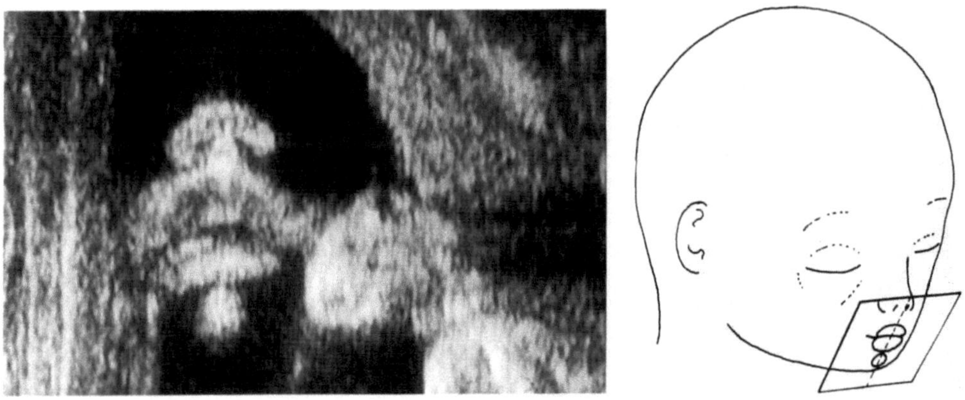

Fig. 4.93 a, b. The nostrils, lips, and chin. As the scan plane is angled lower, less of the nose is visualized, and the chin comes into view

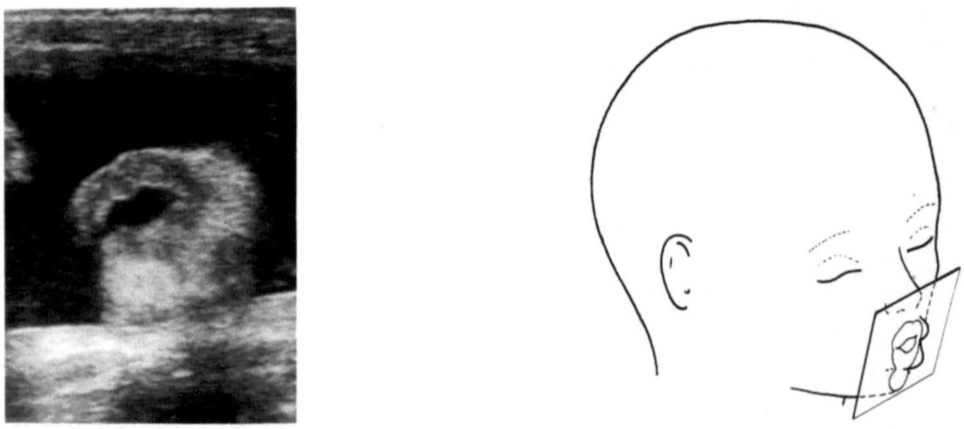

Fig. 4.94 a, b. Scan through the lips and chin with the mouth open and the lips protruding

in this plane. Mayden et al. (1982) also used a frontal scan through the orbits for measuring the biorbital width ("outer orbital diameter") and the interorbital distance ("inner orbital diameter"). Both reference planes are shown schematically in Fig. 4.97 a, b. Measurement in the frontal plane is illustrated in Fig. 4.98 for a 14-week fetus. To ensure an accurate measurement, both scan planes must cut the orbits symmetrically, both orbits must appear equal in size, and the scan must be positioned such that the orbital width is maximal. The biorbital width is measured from the lateral border of the orbit to the opposite lateral border, and the interorbital distance is measured from the medial border of the orbit to the opposite medial border. The biorbital width shows a particularly high correlation with gestational age, and measurements have established its value as a biometric parameter in patients at risk for congenital anomalies. Figure 4.99 shows the nomograms constructed by Jeanty (1984).

4.4.4 Eyes

Moving the transducer anteriorly from the frontal plane of the maximum orbital width often will demonstrate ringlike structures in the area of the globe (Figs. 4.100 and 4.101). Careful analysis will identify these as the four ocular muscles that insert circumferentially around the globe (Fig. 4.101). The lens can be clearly identified only on transverse (Fig. 4.96 a) or paramedian sagittal scans (Fig. 4.102). If the fetal head position is mentoanterior directly beneath the transducer, it is possible to demonstrate the vitreous body, the ocular fundus, and the orbital portion of the optic nerve (Fig. 4.103 a, b). On prolonged observation of this region, the examiner may see movements of the ocular muscle, and the pulsations of the ophthalmic artery may be recorded with a high-resolution transducer. Birnholz (1983) places considerable importance on the evaluation of eye movements for the diagnosis of fetal wellbeing.

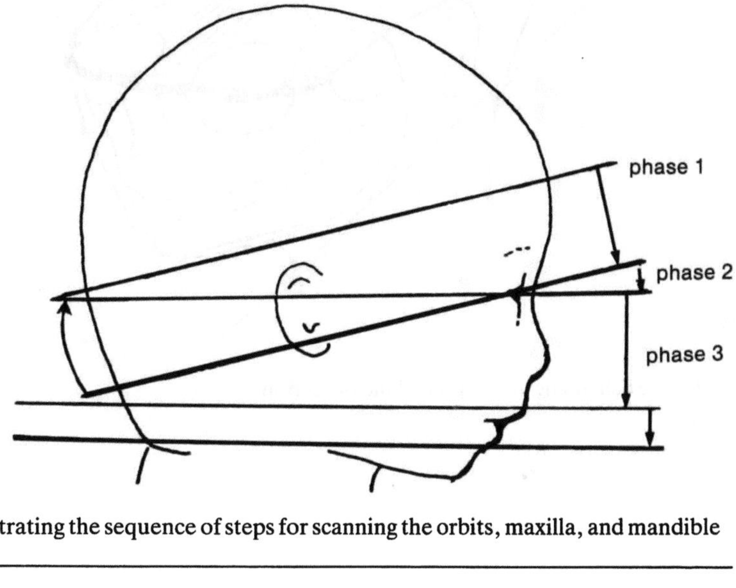

Fig. 4.95. Diagram illustrating the sequence of steps for scanning the orbits, maxilla, and mandible

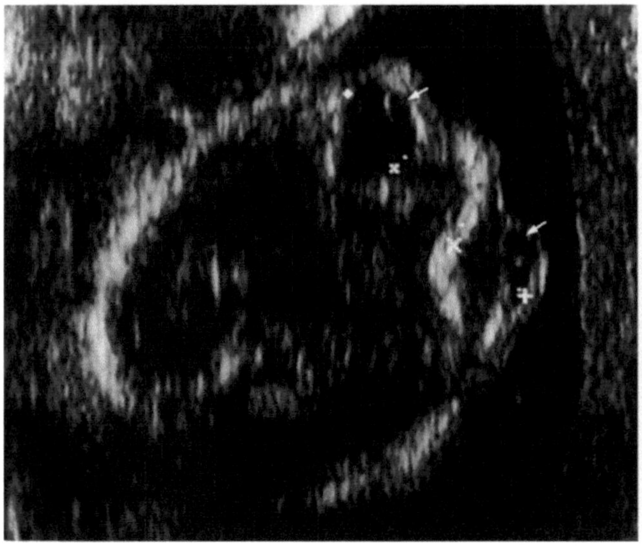

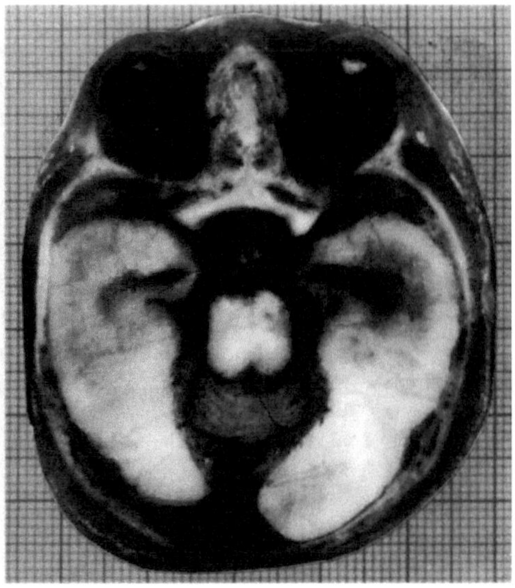

Fig. 4.96. a Transverse scan through fetal skull at 17 weeks in the correct reference plane for measuring the biorbital width and interorbital distance. **b** Analogous frozen section from a 22-week fetus. The lens is visible in the orbit, and the cerebellum is not demonstrated

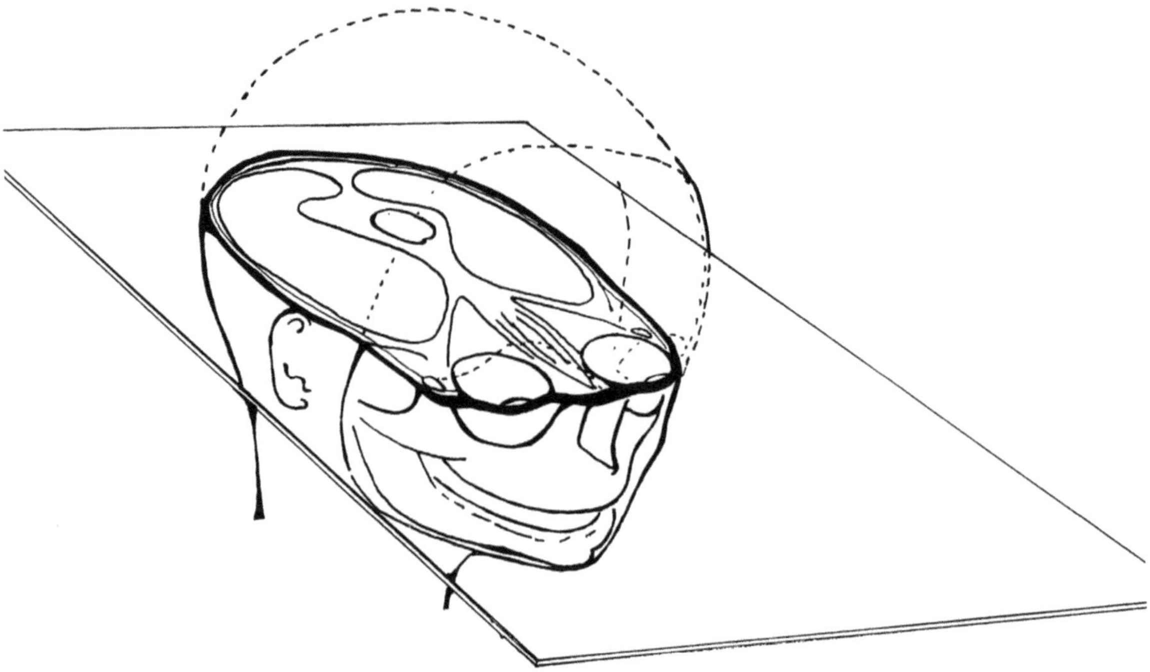

Fig. 4.96. c Schematic view of the plane of section

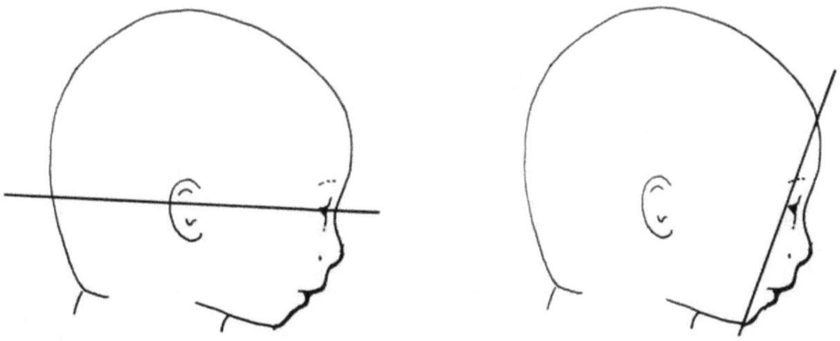

Fig. 4.97 a, b. Diagrams showing the scan planes for the correct measurement of orbital distances. **a** Transverse scan. **b** Frontal scan

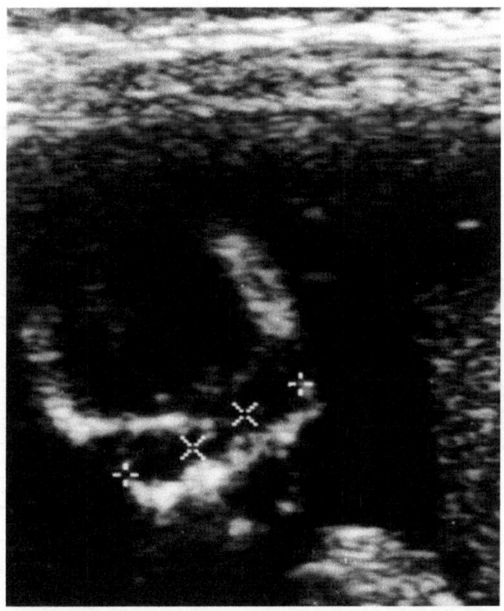

Fig. 4.98. Frontal scan of a 14-week fetus, showing measurement for the orbital distances

The eyelids can be demonstrated by moving the transducer plane anteriorly (Fig. 4.104). They can also be recognized on slightly oblique tangential scans of the facial surface (Fig. 4.105).

4.4.5 Jaw and Tongue

The jaw structures are demonstrated in a "phase 3" maneuver (Fig. 4.95) in which the transducer is moved from the transverse scan through the orbits (Fig. 4.96 a) caudally as far as the tip of the nose (Fig. 4.106). The topographic plane of section is shown in Fig. 4.107. The maxillary arch can be demonstrated as early as 13 weeks (Fig. 4.108). A parallel scan at a slightly more caudal level will delineate the mandible and individual tooth buds (Fig. 4.109). The tongue can be seen between these two sections when the mouth is open (Fig. 4.110).

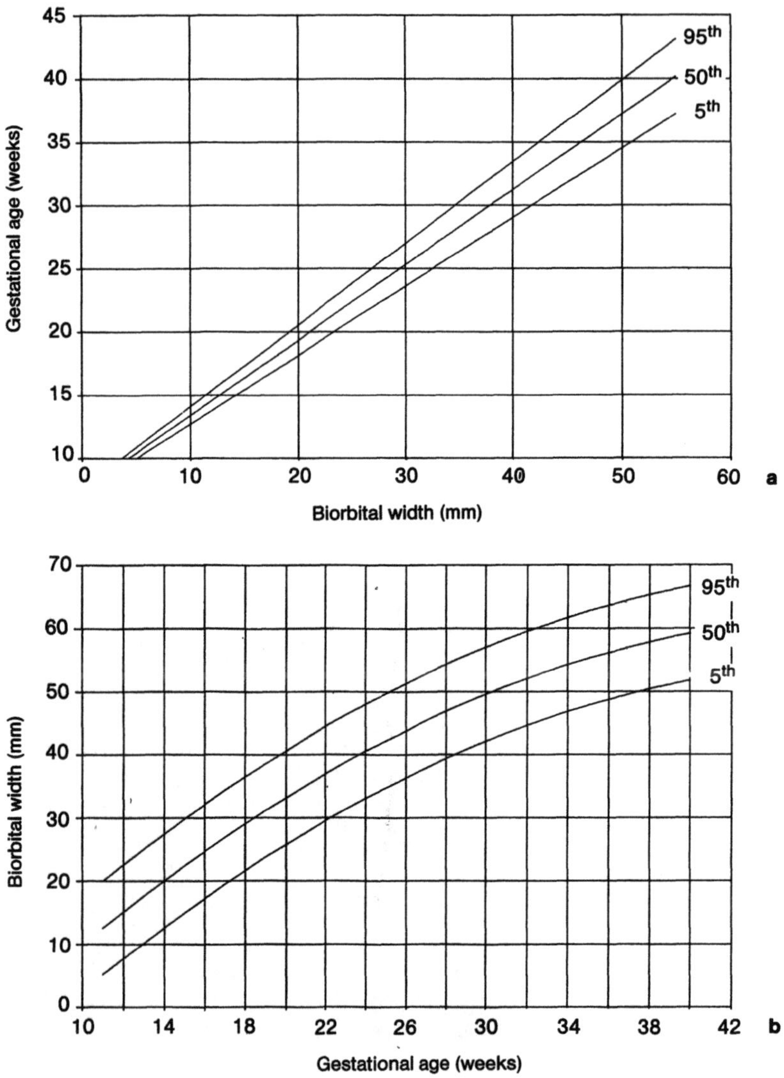

Fig. 4.99. a Graph for the calculation of gestational age from biorbital width. **b** Graph showing relationship of biorbital width to a known gestational age (Jeanty 1986, in Hansmann et al. 1986)

4.4.6 Ear and Neck

The ear can be demonstrated on a scan that is tangential to the lateral aspect of the head (Fig. 4.111). A good image, however, depends on an adequate amniotic fluid volume. Under ideal conditions, tangential scans of the neck region will demonstrate the soft-tissue structures of the shoulders and neck and also the posterior hairline. The hairs will appear highly echogenic when vernix is present (Fig. 4.112).

The importance of phenotype evaluation was discussed in some detail by Hansmann et al. (1985). Smith (1982) published a comprehensive list of facial soft-tissue anomalies and correlates them with specific syndromes and chromosomal defects.

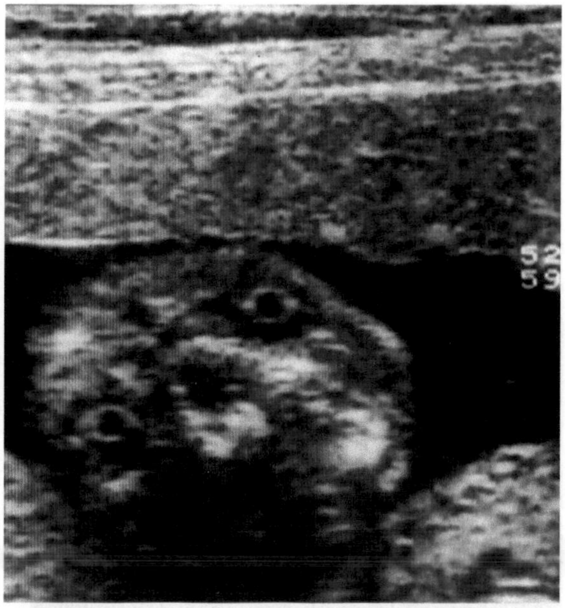

Fig. 4.100. Frontal scan through a fetal head at 20 weeks. The circular contours of the ocular muscles are visible in the orbits

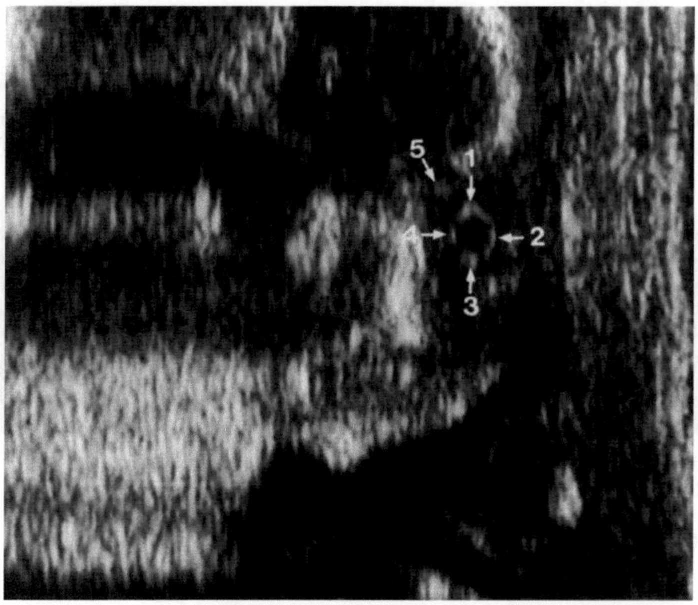

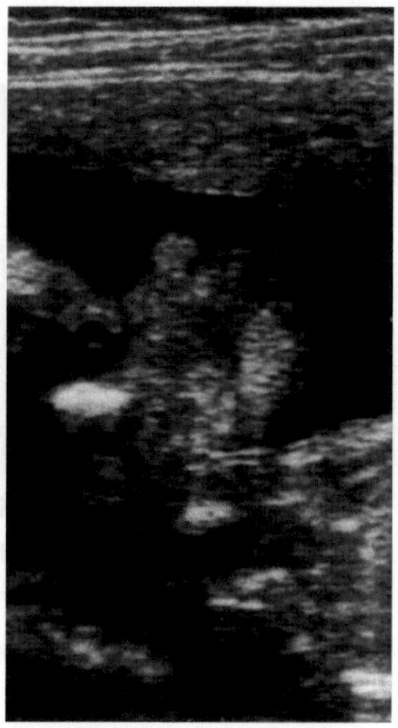

Fig. 4.101. (left) Frontal scan through a fetal head at 16 weeks. The ocular muscles can be identified in the orbit. *1* Superior rectus muscle, *2* lateral rectus muscle, *3* inferior rectus muscle, *4* medial rectus muscle, *5* area of trochlea for guiding the superior obliquus tendon to the medial orbital wall

Fig. 4.102. (right) Paramedian sagittal scan through a fetal face at 21 weeks. The lens appears in the anterior part of the orbit

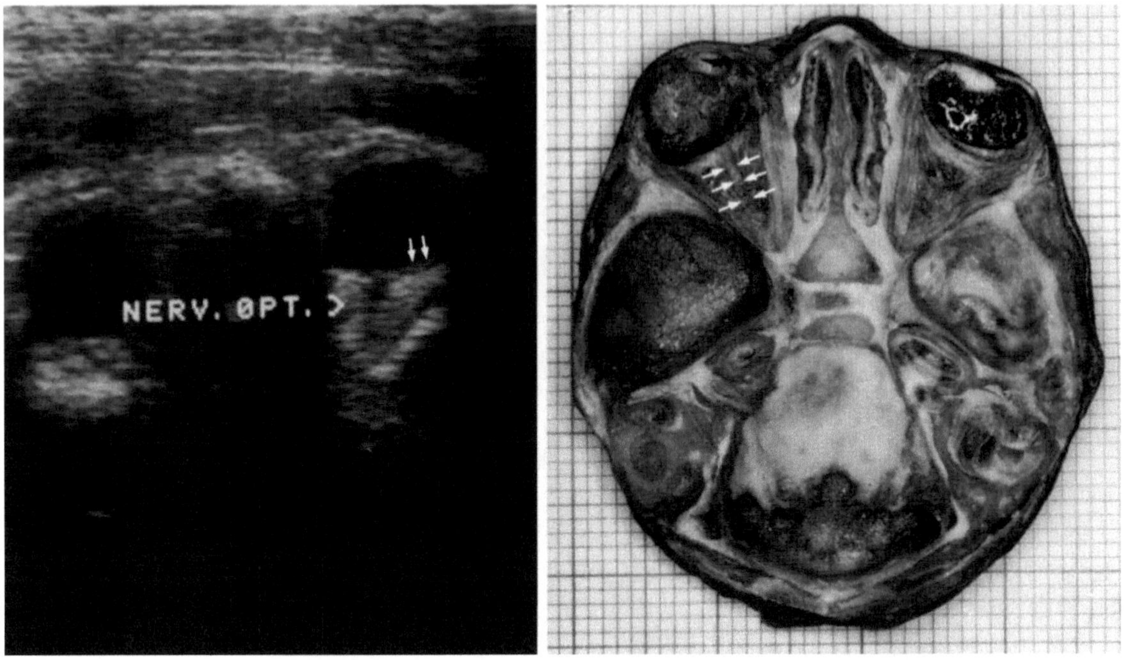

Fig. 4.103. a Transverse scan through a fetal face at 37 weeks. The globes present symmetrically on each side, and the orbital part of the right optic nerve (nerv. opt.) appears as a hypoechoic stripe behind the globe. **b** Analogous frozen section through a fetal head at 15 weeks. *Arrows* mark the orbital part of the optic nerve behind the globe on the left picture

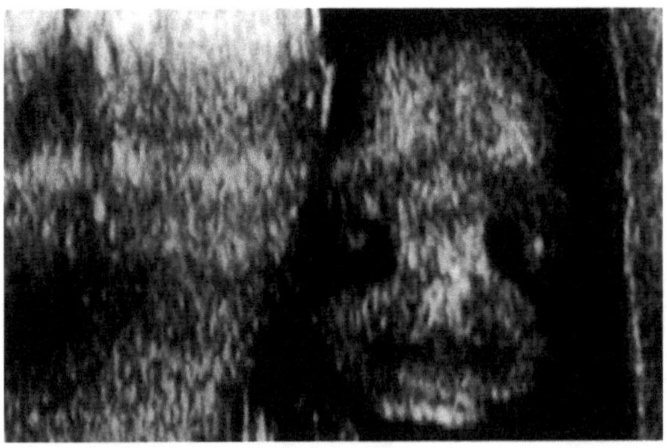

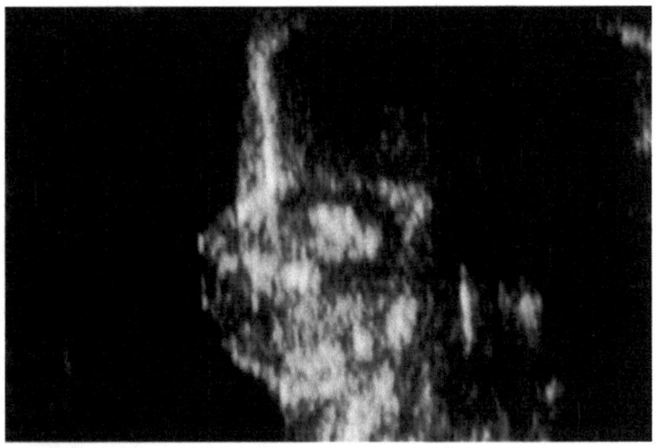

Fig. 4.104. (left) The eyelids are demonstrated by a tangential frontal scan in a 19-week fetus (picture turned 90° for better orientation)

Fig. 4.105. (right) One eyelid is demonstrated on a lateral tangential scan. The fetus is facing left (picture turned 90°)

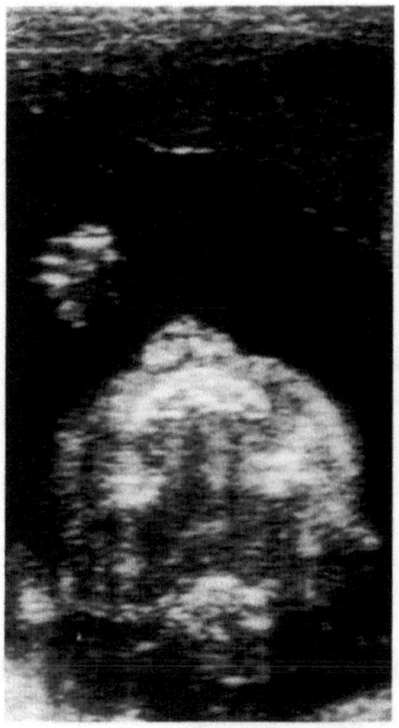

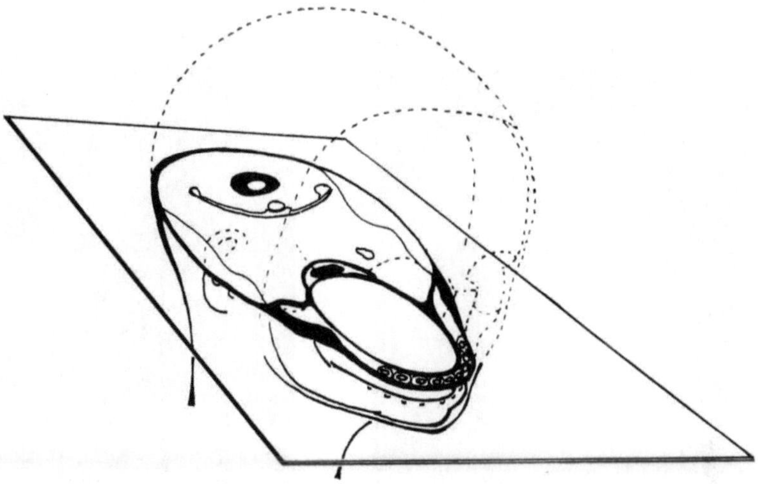

Fig. 4.106. (left) Partial scan through the maxilla in a 19-week fetus. The face is turned upward, and a small portion of the nose is demonstrated as a landmark

Fig. 4.107. (right) Schematic drawing, showing the topography of the scan plane

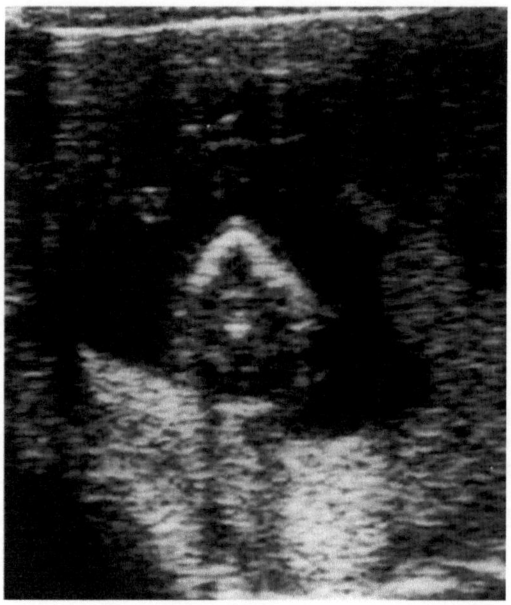

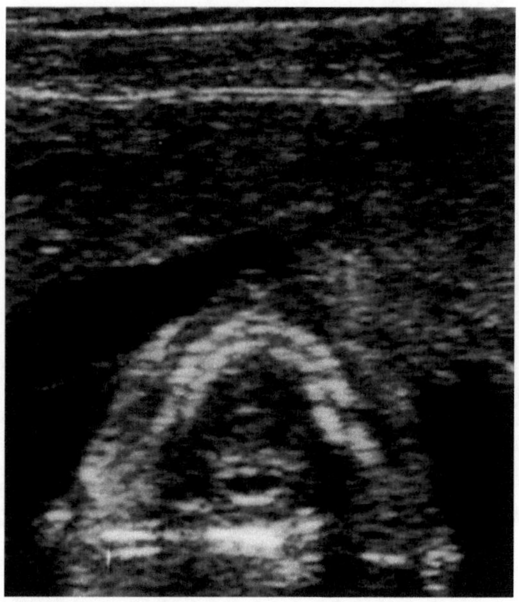

Fig. 4.108. (left) Transverse scan through the maxilla at 13 weeks

Fig. 4.109. (right) Transverse scan through the mandible at 20 weeks. Individual tooth buds are visible on the mandibular arch, and the echo-free space of the oropharynx is seen anterior to the spine

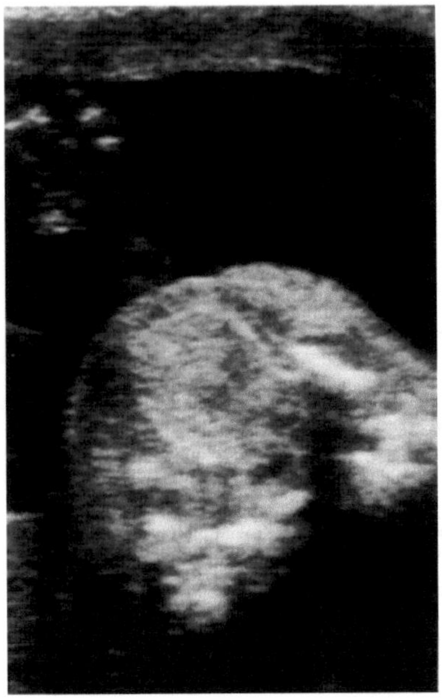

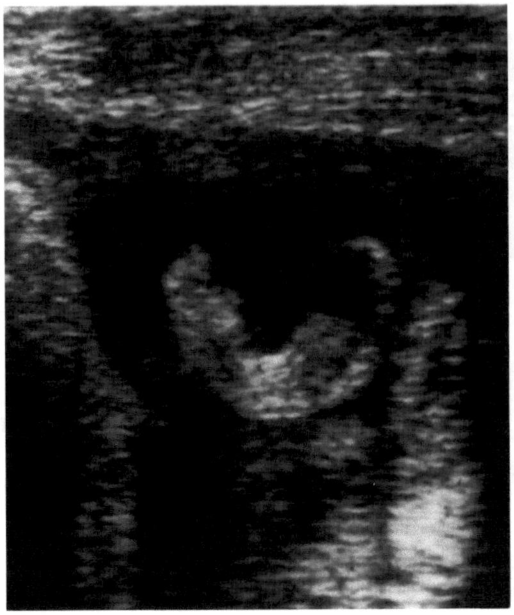

Fig. 4.110. (left) Scan between the maxilla and mandible, demonstrating the tongue (22 weeks)

Fig. 4.111. (right) Tangential scan of the head, demonstrating the ear

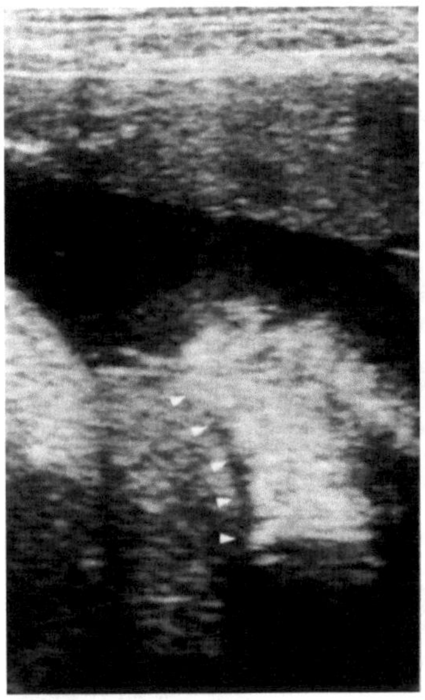

Fig. 4.112. Scan through the neck and shoulder region of a 35-week fetus. The scan shows part of the left shoulder, with the neck region above. The occipital hairs are clearly demonstrated owing to the presence of vernix. The nuchal hairline is defined *(arrows)*

5 Spine and Spinal Cord

Although assessment of the integrity of the fetal spine should be part of any ultrasound examination, it can present significant problems (Miskin et al. 1979; Leucht et al. 1979; Hansmann et al. 1985). The difficulties are illustrated by reports of false-negative diagnoses even at Stage 3 facilities (Hansmann and Gembruch 1984). To perform an adequate sonographic assessment of the fetal spine, it is necessary to understand several basic ontogenic and anatomic concepts.
Chondrification of the vertebrae occurs in the second month and proceeds in a craniocaudal direction, being most pronounced in the vertebral bodies and at the bases of the vertrebral arches. The vertebral arches meet and fuse in the midline in the fourth month, providing dorsal closure of the spinal canal. The peripheral portion of the spinous processes likewise does not appear before the fourth month. Ossification centers begin to form in the vertebrae in the third month. Each vertebra ossifies from three primary centers, one in the body and one in each half of the vertebral arch. From these centers ossification spreads rapidly as gestation proceeds (Fig. 5.1). The earliest point at which the fetal spine could be sonographically detected in our series was at 9 weeks (Fig. 5.2). Evaluation of the spine prior to 15 weeks is pointless given the ontogenic facts outlined above. As in other body areas, examination of the neural tube should follow a standardized procedure that includes scans in sagittal, transverse, and frontal planes. Two facts should be borne in mind:

1. The ossification centers of the vertebrae are always the first spinal structures that can be visualized.
2. To exclude skin defects or lesions of the dorsal aspect of the fetus, a sufficient amount of amniotic fluid must be present between the uterine wall and fetal back, because small cystic structures will not be visualized if the back is pushed against the uterine wall.

It is best to begin the examination by demonstrating the fetal spine on a longitudinal scan.

5.1 Sagittal Sections

For sagittal imaging the transducer should be placed on the maternal abdomen directly above the fetal spine if possible. If the scan is exactly midsagittal and passes between the lateral ossification centers of the vertebral arch, only the vertebral bodies will be demonstrated (Fig. 5.3 a–c). If the scan is angled slightly, it will

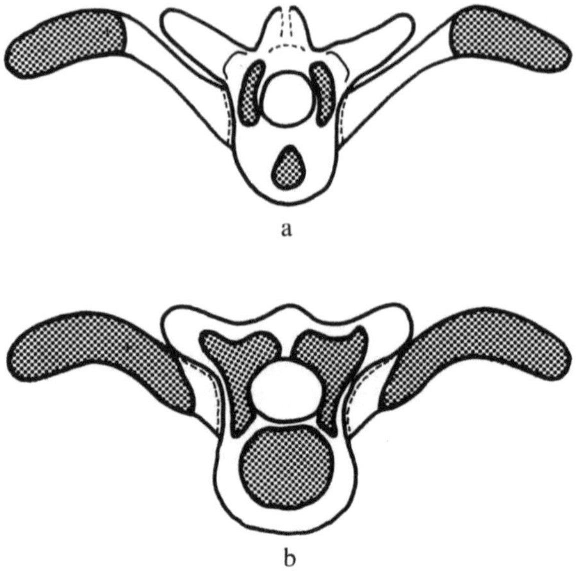

Fig. 5.1 a, b. Schematic drawing of the ossification centers of a thoracic vertebra at 11 weeks (**a**) and at 19 weeks (**b**)

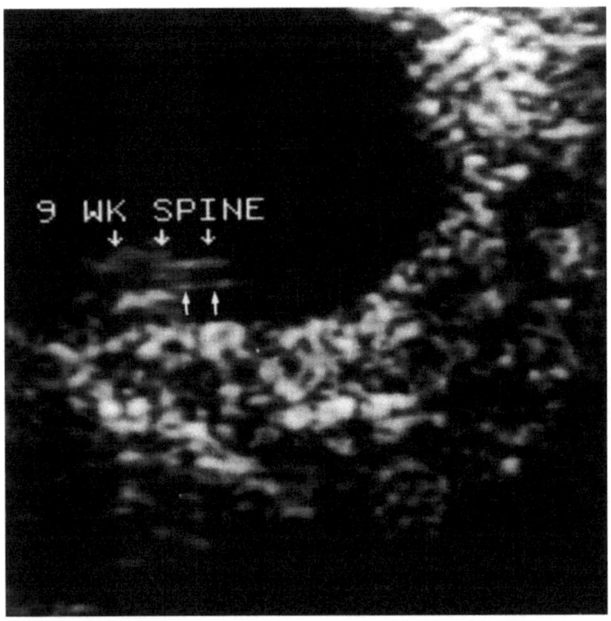

Fig. 5.2. Earliest detection of neural tube contours at 9 weeks. The tangential frontal scan just delineates the double contour of the neural tube

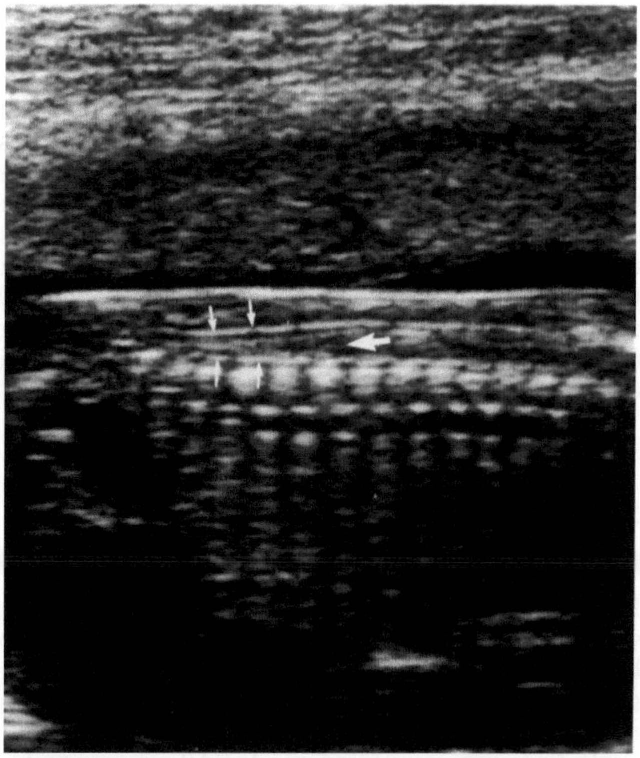

Fig. 5.3. a Exact midsagittal scan through the back of a 19-week fetus. The medullary cone *(horizontal arrow)* lies between the leaves of the spinal dura mater; the ossification centers of the vertebral bodies are seen caudal to it

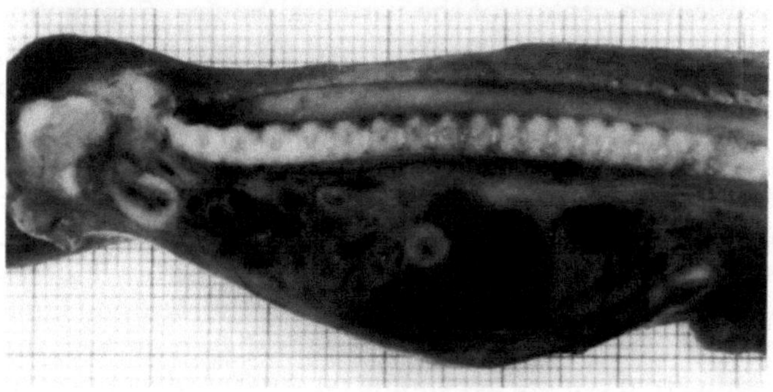

Fig. 5.3. b Analogous frozen section from a 15-week fetus

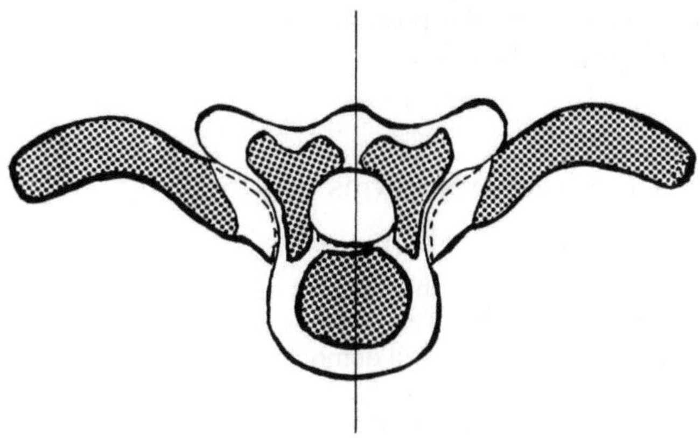

Fig. 5.3. c Schematic drawing of the plane of section through a thoracic vertebra (19 weeks)

create the impression of an intact dorsal bony enclosure of the neural tube. However, the bony structures that appear dorsally are not the spinous processes, but lateral portions of the vertebral arches that have been cut obliquely by the beam (Fig. 5.4 a–c). An exact midline scan will demonstrate the cord and dura mater (Fig. 5.3 a); on an oblique scan, these structures are obscured by acoustic shadows. A small amount of amniotic fluid is present between the placenta and fetal back on both scans. As a result, the dorsal cutaneous surface is clearly outlined on the scans and meningoceles can be excluded. In cases where the fetal back is positioned against the uterine wall or placenta, bilateral pressure on the uterus should suffice to place amniotic fluid between the fetal back and uterine wall. By 19 weeks the outline of the still-unossified coccyx is visible at the caudal end of the spine (Fig. 5.5). A similar scan through the cervical region will demonstrate the medulla oblongata and cisterna magna (Fig. 5.6 a, b).

5.2 Frontal Sections

Following the sagittal examination, the spine usually can be visualized in frontal section by rotating the transducer on the maternal abdomen 90° about the fetal longitudinal axis. Moving the transducer in an anterior-to-posterior direction in this plane will bring various spinal structures into view, according to the depth of the scan and the gestational age of the fetus. As the frontal scan is moved posteriorly from the ventral body surface, the first spinal structures to appear are the ossification centers of the vertebral bodies. Summation of the individual echoes creates the appearance of a linear midline structure (Fig. 5.7 a–d). If the back is arched, the transducer position must be adjusted to the particular area that is being examined. As the scan plane is moved farther posteriorly, differential structural patterns are seen in the thoracic and lumbar regions. In the thoracic spine, an exact frontal scan demonstrates the paired ossification centers of the vertebral arches and, lateral to these, the paired ossification centers of the ribs (Fig. 5.8 a–d). If the spine is straight, the ossification centers of the lateral vertebral arches appear as two parallel lines of echoes in both the thoracic and lumbar areas ("string of beads" pattern, Fig. 5.9). But if the spine is slightly curved, the scan will display different ossification centers depending on the transducer position (Fig. 5.10). In our experience, an extreme posterior scan that is tangential to the fetal dorsal surface will not demonstrate the posterior ossified portion of the vertebral arch prior to 19 weeks (Fig. 5.11 a–b).

5.3 Transverse Sections

The thoracic and lumbar regions of the spine present different structural patterns on transverse scans. The features of the lumbar spine are relatively simple and an exact transverse scan will demonstrate the triangular pattern of the three vertebral ossification centers (Fig. 5.12). The trunk at this level appears circular and symmetrical, and the vertebral body generally can serve as a landmark for correct plane

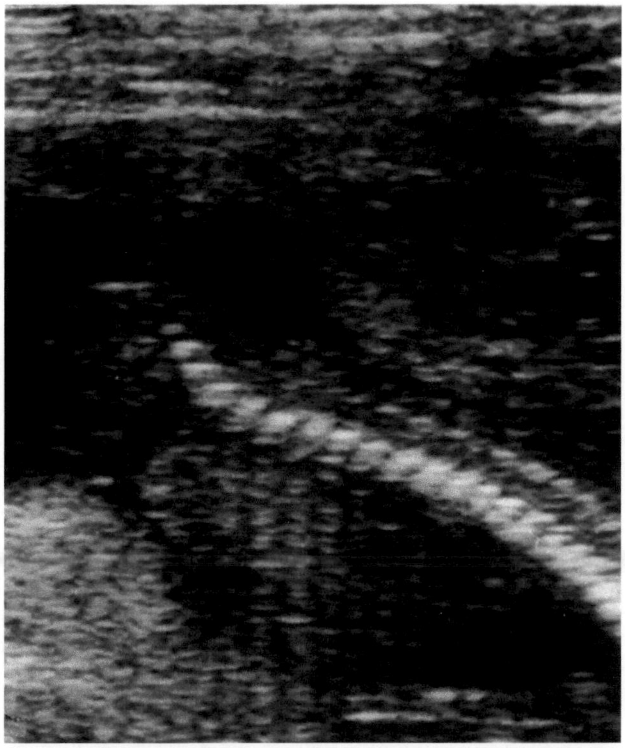

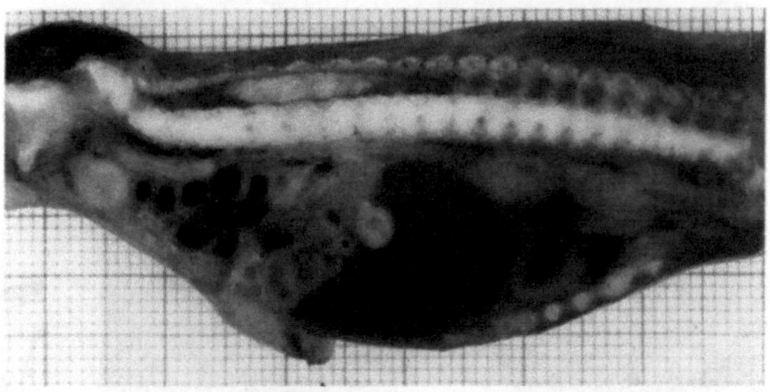

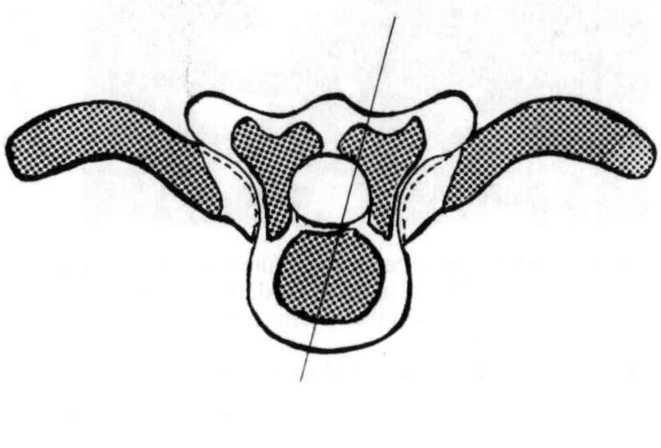

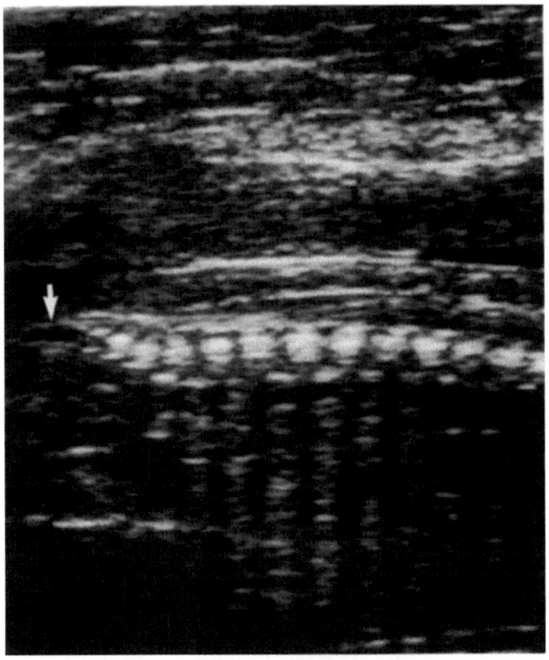

Fig. 5.5. Midsagittal scan through the lumbosacral region of a 21-week fetus. The coccyx *(arrow)* is visible at the caudal end of the sacrum

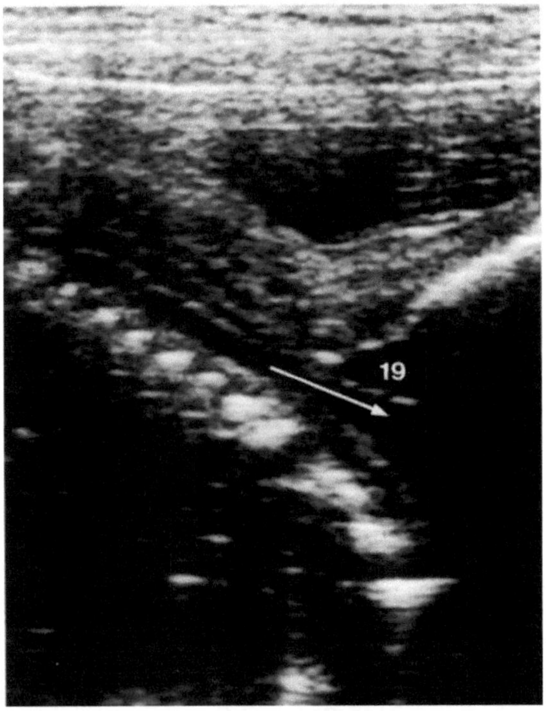

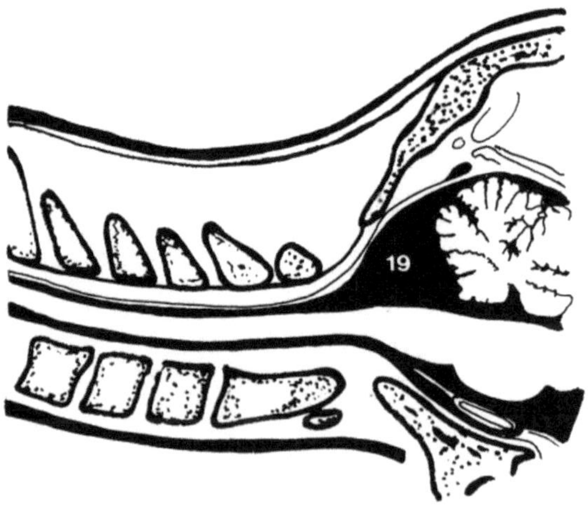

Fig. 5.6. a Midsagittal scan through the neck of a 25-week fetus. The *arrow* indicates the site of entry of the spinal cord into the skull. The tip of the arrow represents the medulla oblongata, behind which a portion of the cisterna magna is visualized *(19)*

Fig. 5.6. b Schematic drawing of the plane of section (*19* cisterna magna)

Fig. 5.7. a Frontal scan through the thoracic vertebral bodies of a 15-week fetus

Fig. 5.7. b Analogous frozen section (15 weeks)

Fig. 5.7. c Schematic drawing of the plane of section

Fig. 5.7. d A frontal section at this level traverses only the vertebral bodies

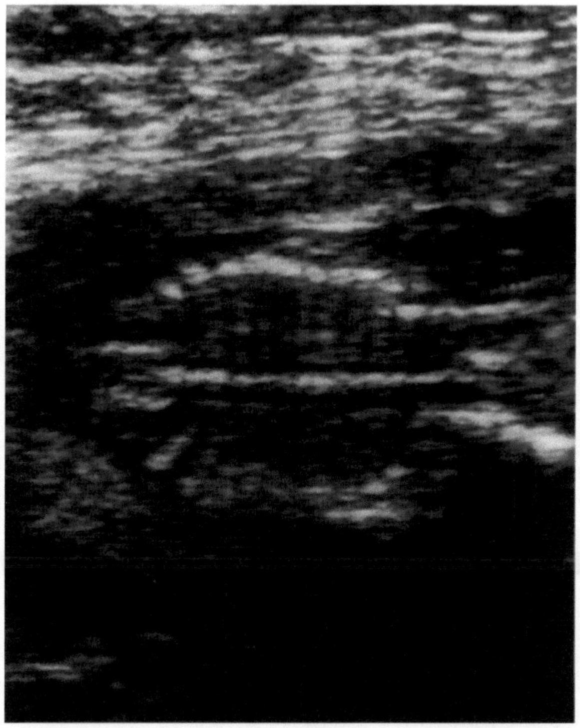

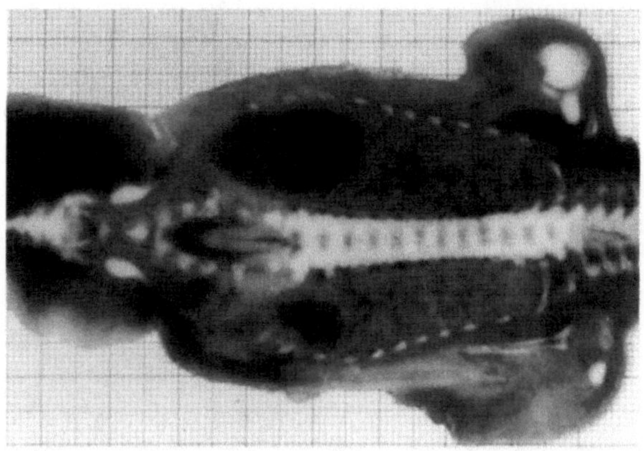

b

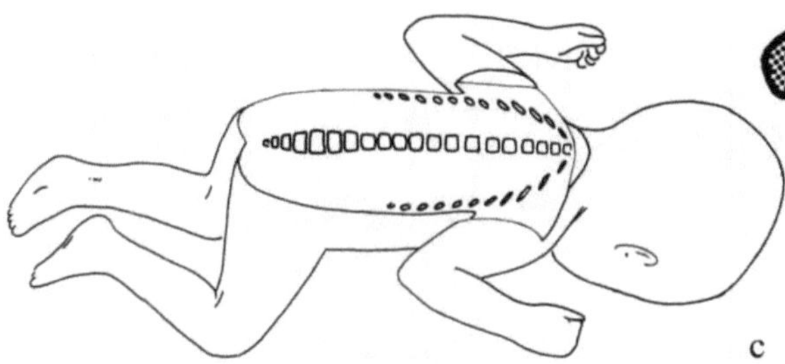

c

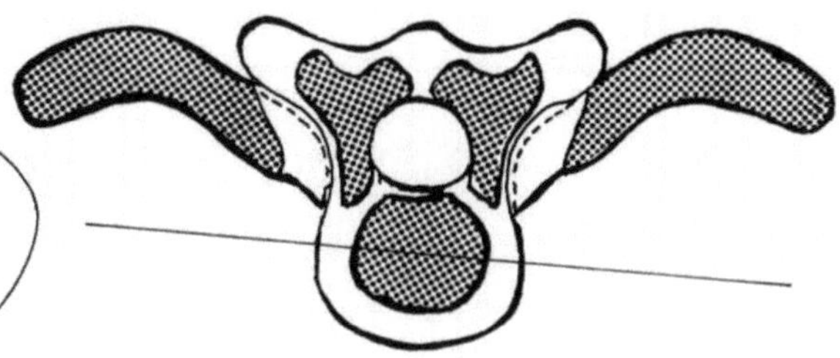

d

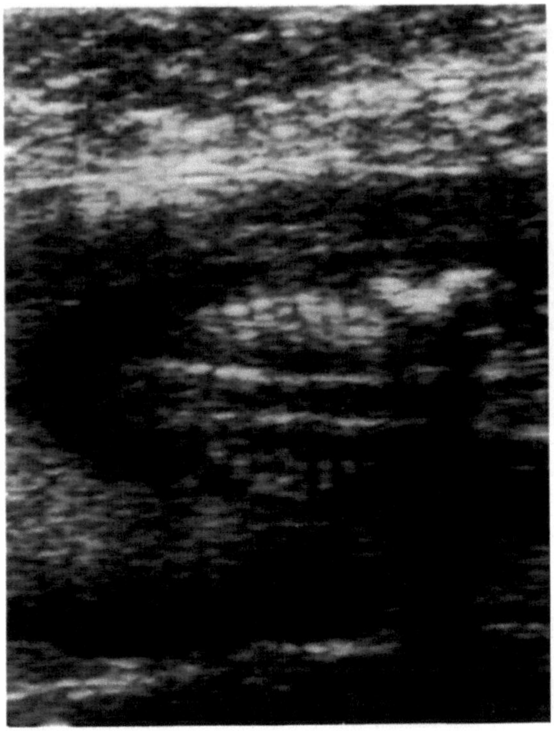
a

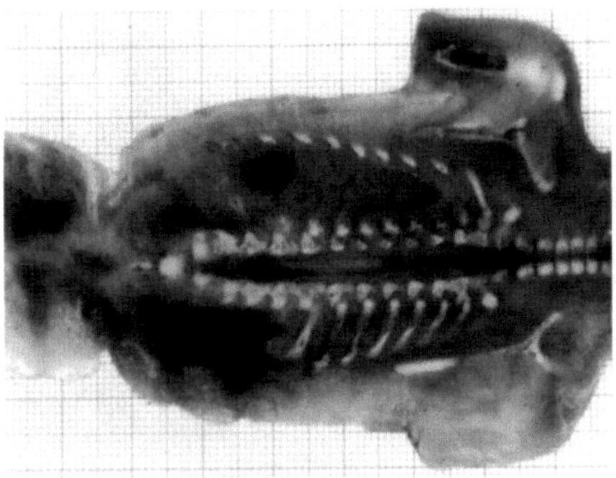
b

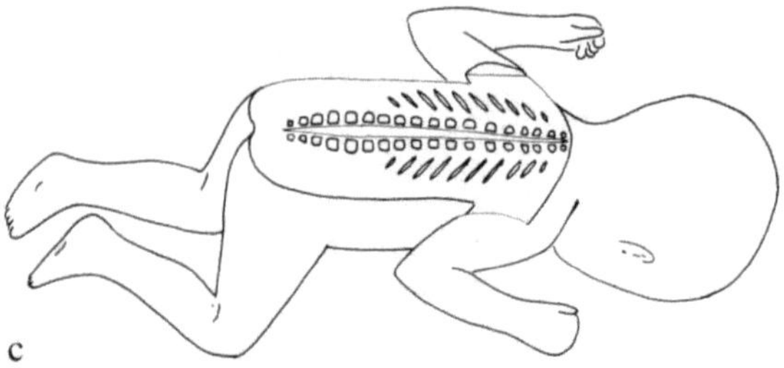
c

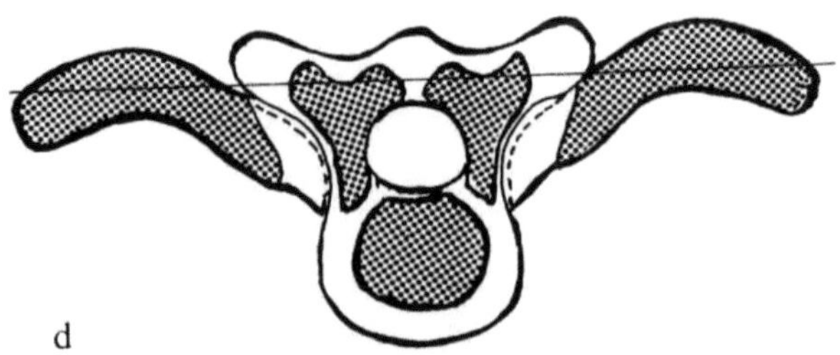
d

Fig. 5.8. a Frontal scan through the thoracic spine of a 15-week fetus. The beam cuts the ossification centers of the vertebral arches and ribs

Fig. 5.8. b Analogous frozen section from a 15-week fetus

Fig. 5.8. c Schematic drawing of the entire plane of section

Fig. 5.8. d Schematic drawing of the plane of section through a thoracic vertebra

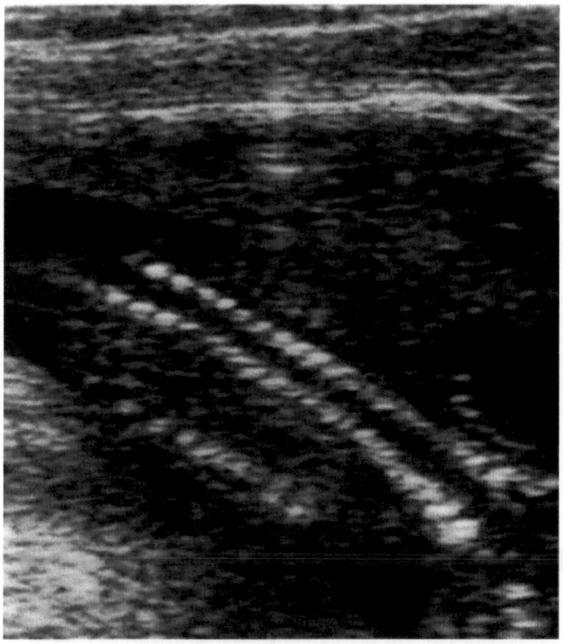

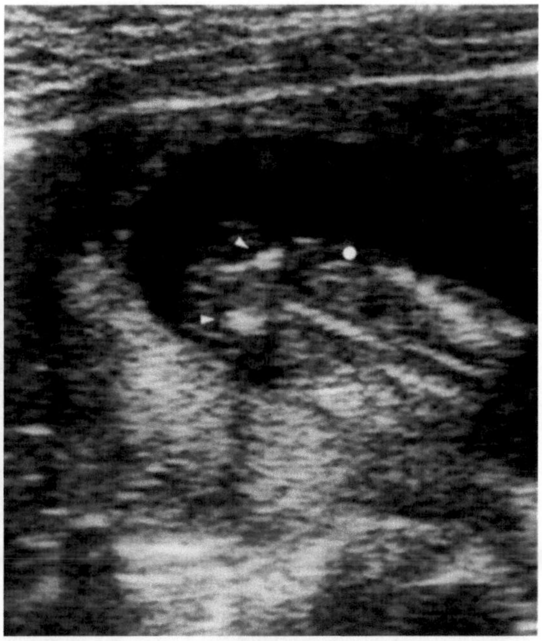

Fig. 5.9. (left) Frontal scan demonstrating all lateral spinal ossification centers in a 14-week fetus

Fig. 5.10. (right) Frontal scan through the back of a 13-week fetus. This scan cuts the ossification centers of the vertebral arches in the thoracic spine, and the ossification centers of the vertebral bodies in the lumbar spine. The *arrows* mark the ilium. The *dot* is over the left kidney

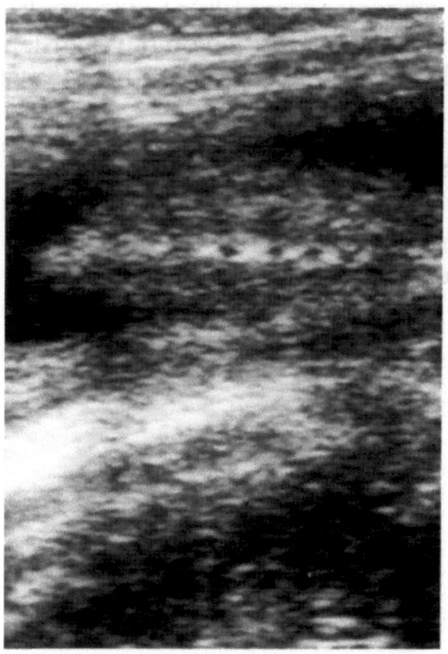

a

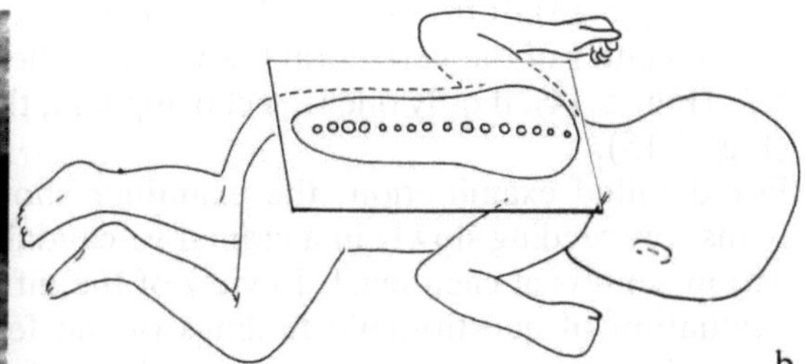

b

Fig. 5.11. a Tangential frontal scan through the back of a 21-week fetus. The tips of the spinous processes appear centrally between the dorsal soft tissues

Fig. 5.11. b Schematic drawing of the plane of section

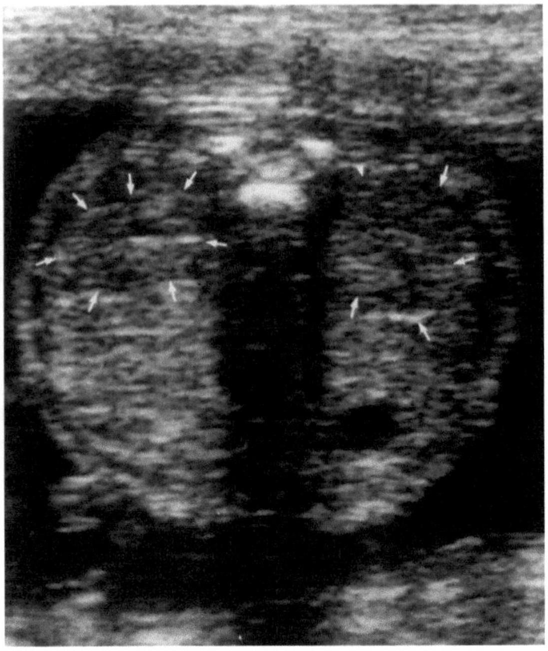

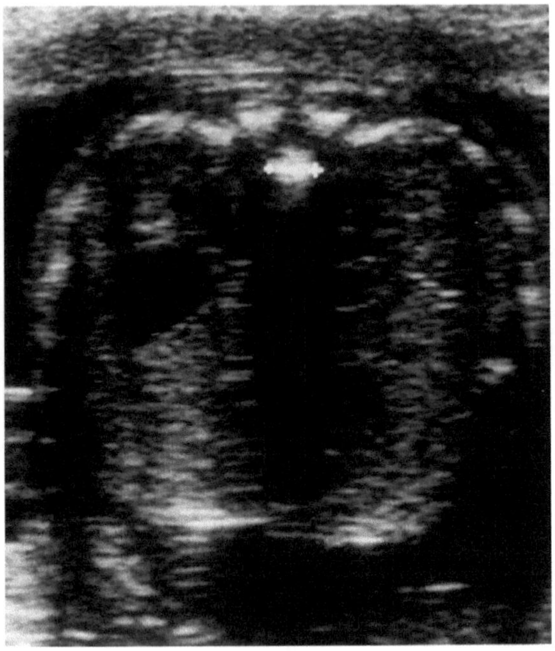

Fig. 5.12. (left) Transverse scan through the region of the lumbar spine. Note the three ossification centers of the lumbar vertebra. The kidneys flank the spine *(arrows)*

Fig. 5.13. (right) Exact transverse scan through the region of the thoracic spine. The three ossification centers of the vertebral body and arches are flanked at this level by the ossification centers of the ribs. The cursors mark the transverse dimension of the vertebral body (23 weeks, 4 mm)

selection. In the region of the thoracic spine, an exact transverse scan is signified by the three vertebral ossification centers and by the paired ossification centers of the ribs (Fig. 5.13). If the ribs and lateral ossification centers are visible but the ossification center of the vertrebral body is not, the scan plane is tilted on the right-left axis (Fig. 5.14); if only one rib echo appears, the scan plane is tilted on the AP axis (Fig. 5.15).

For detailed examination, the examiner should perform a series of transverse scans, proceeding slowly in a cranial-to-caudal direction and examining the ossification centers at each level. In view of the difficulties mentioned above, the final evaluation of questionable findings of the fetal spine generally should be performed at centers with an appropriate level of expertise. Figure 5.16 illustrates the various types and degrees of neural tube defect that can occur (Rickham et al. 1975) in order to facilitate their recognition.

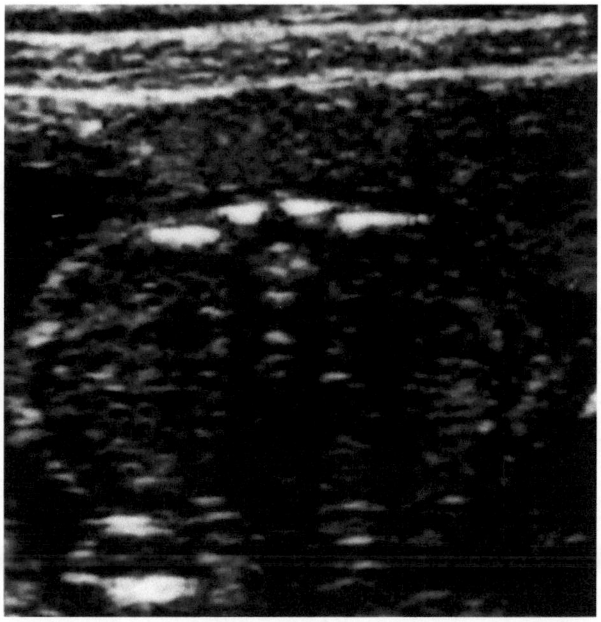

Fig. 5.14. This transverse scan through the thoracic spine is tilted on the right-left axis. Only the ossification centers of the vertebral arches and ribs are visualized

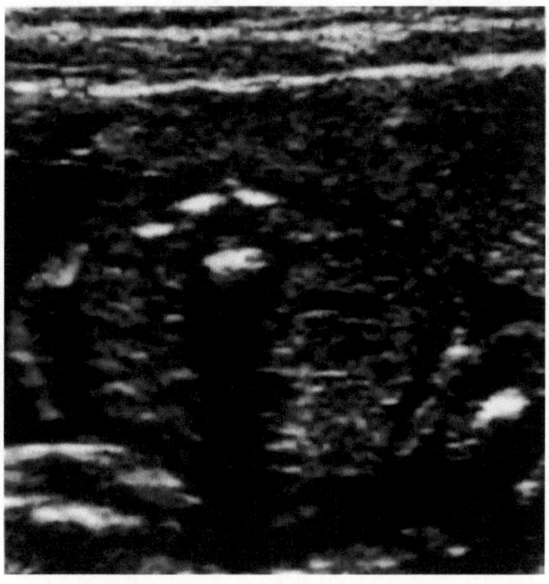

Fig. 5.15. This scan is tilted slightly on the AP axis and cuts the ossification center of only one rib

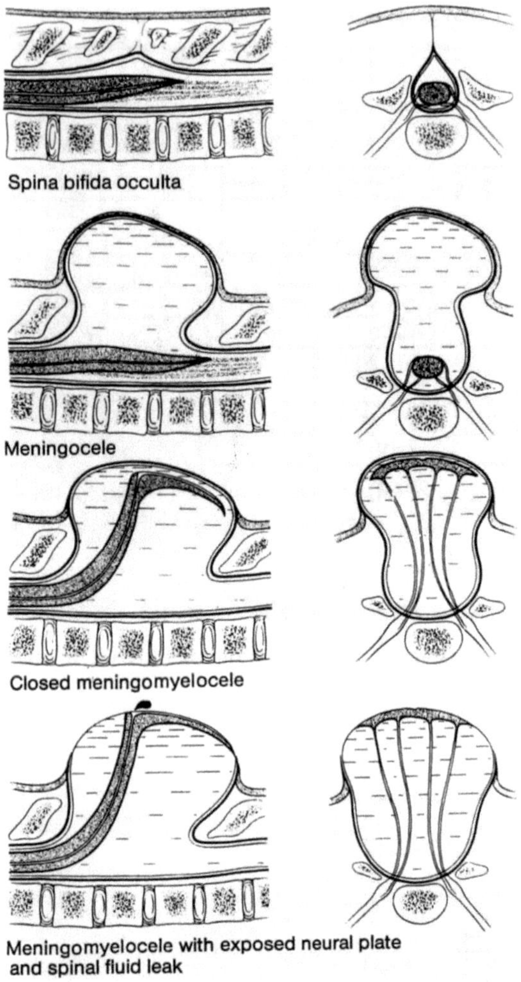

Fig. 5.16. Schematic drawings illustrating the grades of severity of neutral tube defects (Rickham et al. 1975)

6 Neck

A separate chapter is devoted to the neck on anatomic grounds. Very little information has been published to date on the normal anatomy of the fetal cervical region (Jeanty et al. 1984; Cooper et al. 1985).

In our experience, the structures of diagnostic importance in the cervical region–the pharynx, trachea, and great vessels–can be delineated by ultrasound. Portions of the pharynx frequently appear on midsagittal profile scans of the fetal head (Fig. 6.1). The cranial and caudal portion of the pharynx usually are not visualized due to shadowing from the maxilla and mandible, and only the oropharynx can be appreciated behind the tongue. The oropharynx can also be seen on transverse scans through the tongue where the beam is directed between the maxilla and mandible (Fig. 6.2). The trachea can be visualized by moving the transducer caudally over the neck, assuming the fetal head is not flexed forward (Fig. 6.3). The trachea characteristically appears as a hypoechoic band like structure whose borders contrast sharply with the brighter echoes from the cartilages (Figs. 6.3 and 6.4). Cooper et al. (1985) could demonstrate the trachea in 94% of cases in a selected population between 17 and 38 weeks. Measurements showed that the tracheal diameters increased slightly from 2.4 mm at 17–25 weeks to 2.8 mm at 25–38 weeks. Because the instruments we used were not calibrated for less than one millimeter, we were unable to confirm these findings. The trachea can also be identified on transverse scans through the caudal neck region, especially if the section includes one of the cricoid cartilages. One such scan is shown in Fig. 6.5, where thyroid tissue overlies the trachea anteriorly.

On frontal scans, the trachea is delineated by the characteristic appearance of the echogenic cartilages (Fig. 6.6 a). The laryngopharynx can be recognized on frontal scans cranial to the trachea during swallowing and breathing movements (Fig. 6.6 a–c). The fluid content of this portion of the pharynx can be seen to vary markedly with the individual swallowing and respiratory phases. Portions of the epiglottis may appear on frontal scans at the junction of the trachea and laryngopharynx. On such sections the piriform recess appears to the left and right of the epiglottis. From 17 weeks we were able to observe rhythmic changes in diameter, which tended to be associated with tongue movements. Utsu et al. (1983) used Doppler measurements to study tracheal fluid flow associated with breathing movements.

We were unable to demonstrate the esophagus in any case, although in several instances we did observe transient filling with fluid and peristalsis. This process was always accompanied by a retractile movement of the tongue and a collapse of the oropharynx. Observations to date are insufficient to permit a detailed evaluation of fetal swallowing.

There are already reports on the possibility of diagnosing esophageal atresia by ultrasound (Farrant 1980; Bowie and Clair 1982; Pretorius et al. 1983; Eyheremendy and Pfister 1983), but further confirmatory studies are needed.

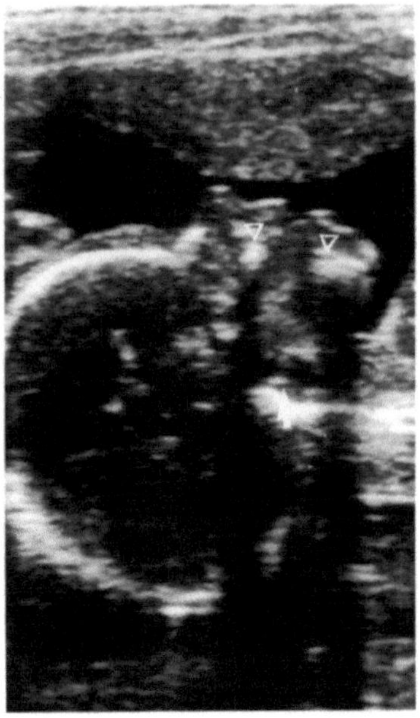

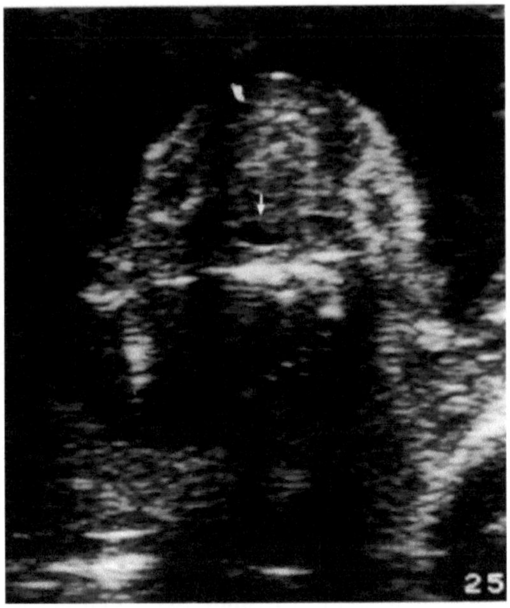

Fig. 6.1. (left) Exact midsagittal scan through the fetal skull at 19 weeks. The orpharynx is visible behind the tongue *(arrow)*; acoustic shadows from the mandible and maxilla *(arrows)* obscure the cranial and caudal portion of the pharynx

Fig. 6.2. (right) Transverse scan through the fetal head at 24 weeks. The mouth is open, and the oropharynx *(arrow)* is visible behind the tongue

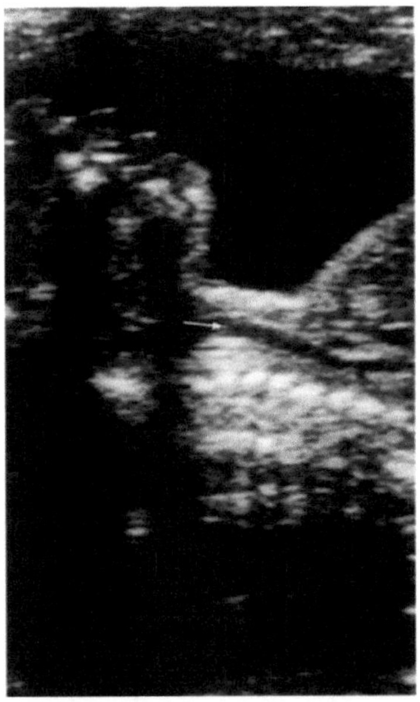

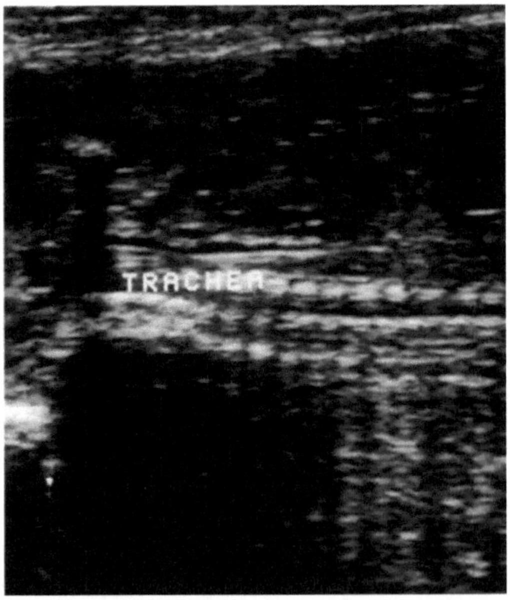

Fig. 6.3. (left) Midsagittal scan through a fetal head at 19 weeks. The head is extended, and the trachea *(arrow)* is visible within the neck

Fig. 6.4. (right) Sagittal scan through the neck and thorax of a 20-week fetus (part of the head appears on the left side of the image). The trachea lies anterior to the spine and is visible in its full length

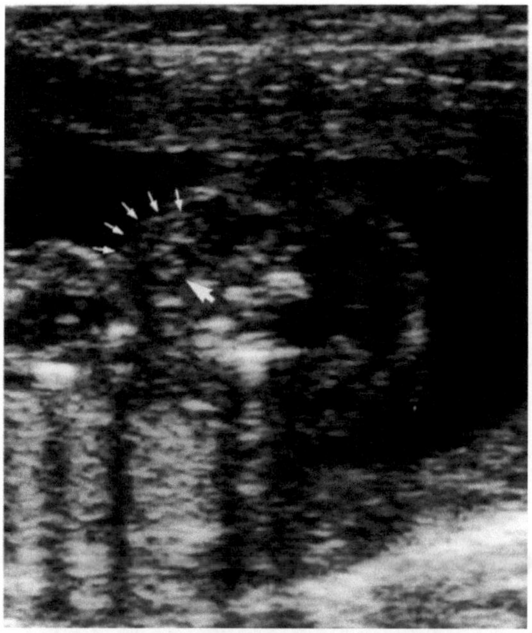

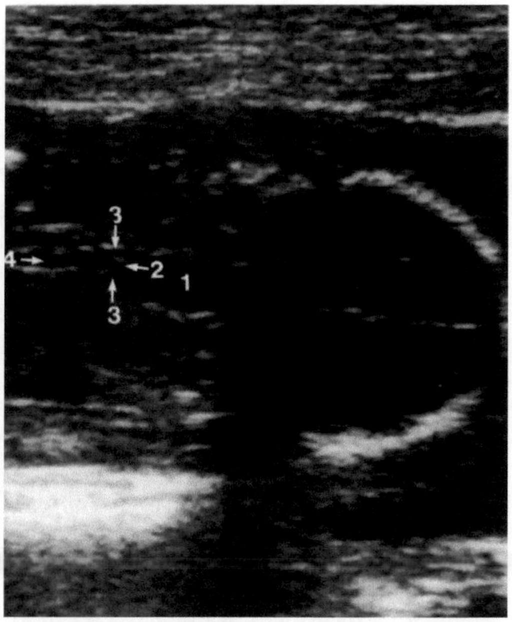

Fig. 6.5. (left) Transverse scan through the neck of a 20-week fetus. The thyroid gland *(small arrows)* presents anterior to the trachea *(large arrow)*

Fig. 6.6. a (right) Frontal scan through the head and neck of an 18-week fetus (head to the right) at the level of the laryngopharynx (*1*), caudal to which is the epiglottis (*2*), demarcated on both sides by the piriform recess (*3*). The trachea (*4*) appears farther caudally

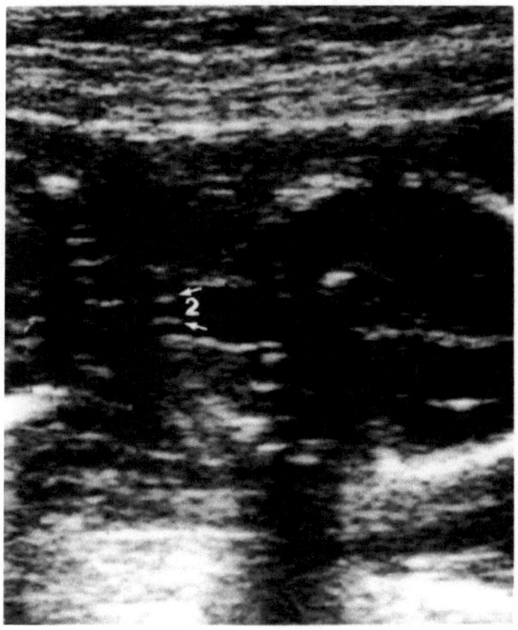

b

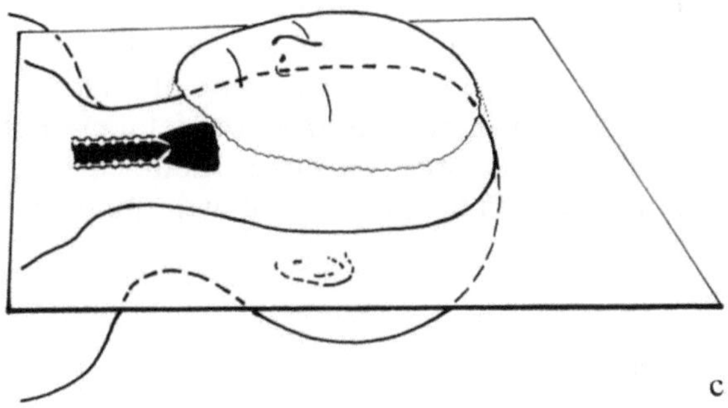

c

Fig. 6.6. b Analogous scan plane in the same fetus during swallowing. Note the expansion of the pharynx. The epiglottis (*2*) is imaged on a slightly different plane than in the previous figure

Fig. 6.6. c Schematic drawing of the image plane

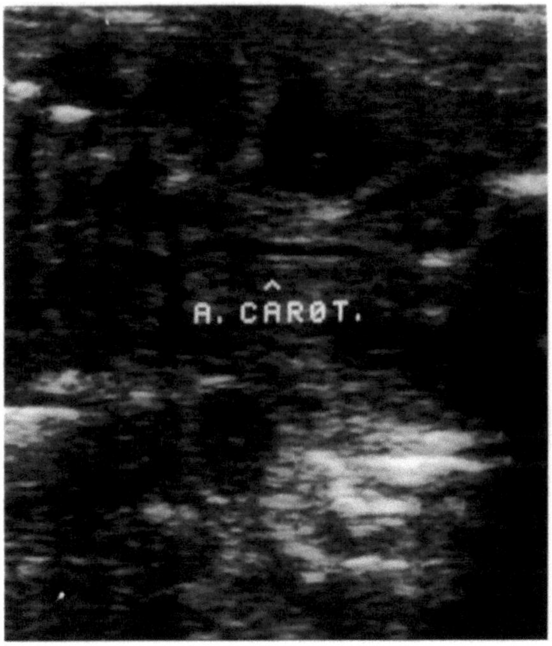

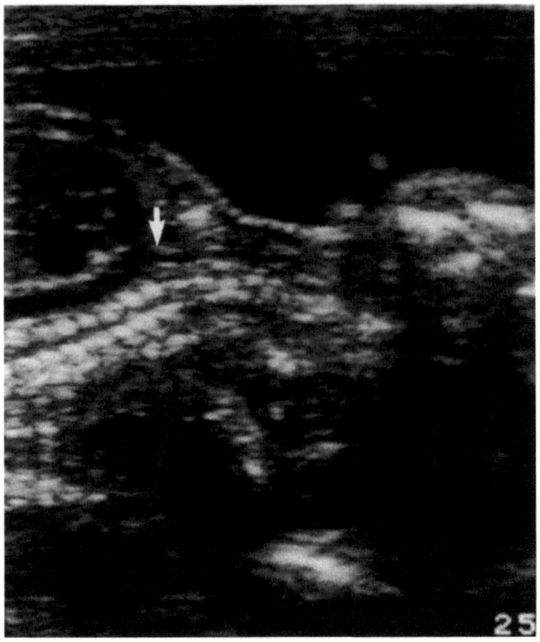

Fig. 6.7. (left) Frontal scan through the neck of a 23-week fetus. The head is at the right side of the image. The scan displays the common carotid artery in the neck

Fig. 6.8. (right) Paramedian sagittal scan through the neck and thorax of a 24-week fetus. The origin of the left common carotid artery *(arrow)* is demonstrated

The common carotid arteries (Fig. 6.7) are very clearly seen in the lateral cervical region, especially on frontal scans, and are easily recognized by their pulsations. If doubt exists, the artery can be identified by tracing it to its origin from the aorta (Fig. 6.8). Generally we do not include the fetal neck in a basic examination, because flexion of the cervical spine often interferes with selective visualization of that region, or shadowing from the fetal extremities prevents accurate identification of the structures. However, detailed examination of the cervical region is recommended in the presence of hydramnios for the reasons stated earlier.

7 Thorax (Heart, Lung, Great Vessels)

The illustrations in this chapter are labeled with letters and numbers that are explained in Table 7.1. The dominant intrathoracic organ is the heart.

The anatomy and spatial orientation of the fetal heart make it one of the most difficult organs to evaluate by ultrasound, even at a Stage 2 or Stage 3 facility. This is due to the heart and its axes being oriented at an angle to the standard body axes, and to the varying directions and crossing of the blood vessels emerging from the heart. Finally there is the problem of fetal position, which determines the ability to visualize cardiac structures and which cannot be influenced by the examiner. Because of the early ossification of the ribs (see Chap. 10), the ultrasound windows to the fetal heart are less than optimal.

In pediatric and adult two-dimensional echocardiography, it is emphasized that a qualitatively and quantitatively accurate evaluation relies on a standardized examination procedure and the reproducible documentation of findings (Grube 1985). Considering the fact that all known echocardiographic approaches to the pediatric and adult heart are identified from well-defined anatomic landmarks on the body surface, and that most examiners use sector transducers that can be applied directly over the area of interest, it is little wonder that fetal echocardiography tends to yield substantially less diagnostic information. Experience in recent years has shown that an adequate examination of the fetal heart can be performed only by a specially trained examiner who is familiar with the techniques of two-dimensional, M-mode, and Doppler echocardiography (Redel and Hansmann 1984).

The diagnostic capabilities of the combined application of these techniques and the associated problems have been widely discussed (Allan et al. 1980; Sahn et al. 1980; Wladimiroff 1981; Wladimiroff and McGhie 1981; Jeffrey 1982; DeVore et al. 1982; Redel and Hansmann 1984; Redel et al. 1984a, b; Nisand et al. 1984).

Because the avove-mentioned techniques are frequently not available for routine or even Stage 1 and Stage 2 examinations, there is a need to evaluate how much

Table 7.1. List of abbreviations used for intrathoracic organs and blood vessels

Right atrium	a	Superior vena cava	1	Foramen ovale	→
Left atrium	b	Ascending aorta	2	Septum primum	S_1
Right ventricle	A	Descending aorta	2d	Myocardium	M
Left ventricle	B	Pulmonary trunk	3	Papillary muscle	mp
Ventricular septum	Vs	Ductus arteriosus	4	Lung	P
interatrial septum	Sa	Pulmonary arteries	5	Trachea	T
Tricuspid valve	vt	Inferior vena cava	6	Bronchus	B
Mitral valve	vm	Pulmonary veins	7	Esophagus	O
		Azygos vein	8		

can be accomplished with real-time equipment alone in fetal cardiac scanning. Because the diagnosis of cardiac anomalies has been described using basic real-time ultrasound equipment (Winter et al. 1979; Köhler et al. 1981; Hansmann et al. 1985), it is appropriate to review the sonoanatomic features of the fetal heart in this volume.

It is assumed that the reader knows the essentials of fetal cardiac anatomy, so they will not be repeated here. To aid orientation, Fig. 7.1 shows the topographic relations of the great vessels in a heart from a 22-week fetus. In Fig. 7.1 a, a probe has been inserted through the superior vena cava and right atrium into the inferior vena cava; in Fig. 7.1 b, the heart has been rotated forward to show the relationship of the ductus arteriosus to the aorta. In both cases the trachea has been left attached to the specimen.

7.1 "Four-Chamber View"–Plane of Section

For fetal cardiac evaluation during basic examination, the "four-chamber view" is essential. An adequate four-chamber view can be obtained from 15 weeks according to Nisand et al. (1984) and from 19 weeks according to Redel and Hansmann (1984).

Figure 7.2 a–c shows the four-chamber view as it appears in the sonogram, in frozen section, and diagramatically. The sonogram was rotated 180° so that all the pictures would conform to the conventional anatomic presentation. To allow for the roughly transverse orientation of the fetal heart relative to the trunk, and to avoid shadowing from the ossification centers of the vertebral bodies and ribs, the best four-chamber view is obtained by directing the beam from the lower left side of the heart to the upper right, as shown in Fig. 7.2 c. The relationship of the beam plane to the fetal thorax is shown in Fig. 7.3.

7.2 Orientation in the Four-Chamber View

To interpret the four-chamber view correctly, we must know the position of the observer in relation to the fetal presentation. Otherwise we cannot tell the right side of the image from the left. There are basically two ways in which the scan image can be displayed:

1. The observer is looking down on a slightly oblique transverse section through the thorax (Fig. 7.4 a–c). The fetal spine is at 7 o'clock. When the section is viewed in this manner, the aorta presents slightly lateral and to the left of the spine owing to its asymmetrical position. The cardiac apex points anteriorly and to the left. The four-chamber view displays both ventricles, the ventricular septum, both atria, and the foramen ovale.
2. In Fig. 7.5 a–c, the image is displayed as though the observer were viewing the section from below. In this presentation the aorta appears slightly to the right of the spine, and the cardiac apex points toward the right side of the image. The orientation of the four-chamber image will vary with the general examination setup and with the presentation and position of the fetus.

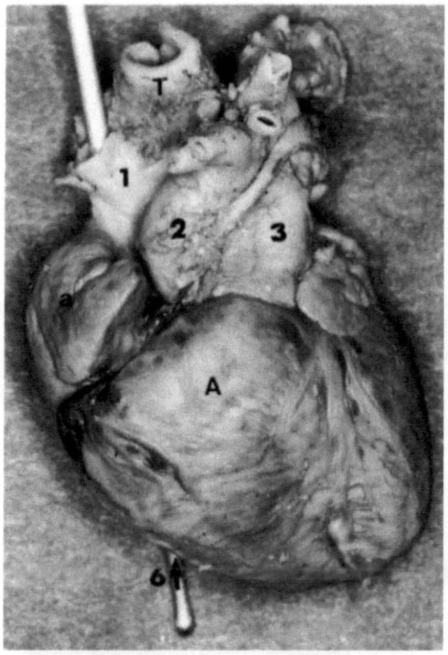

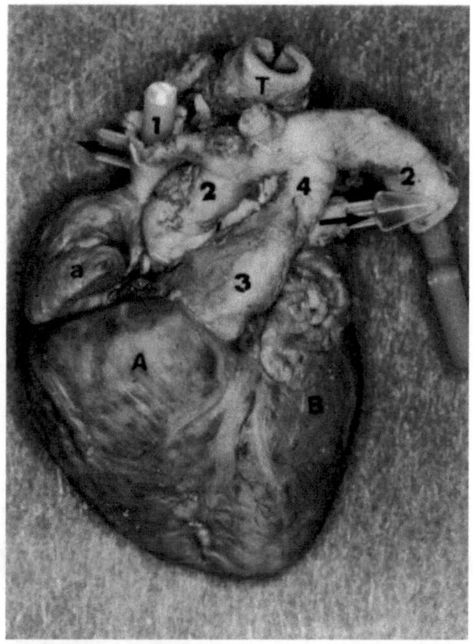

Fig. 7.1. a Fetal heart at 22 weeks, positioned to show its orientation within the thorax. The probe marks the superior and inferior vena cava

Fig. 7.1. b The left side of the heart has been turned forward to reveal the ductus arteriosus. The descending aorta and pulmonary veins are marked with *arrows*. A probe has been inserted into the superior vena cava (*1*)

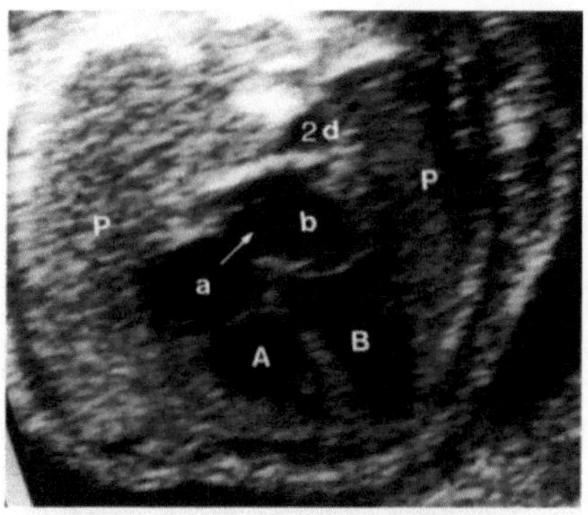

a

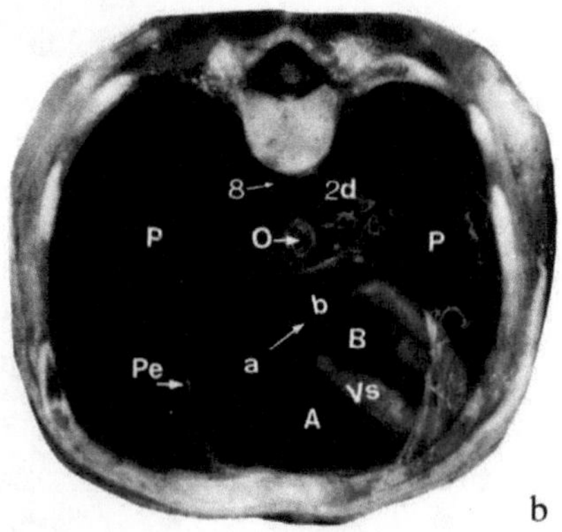

b

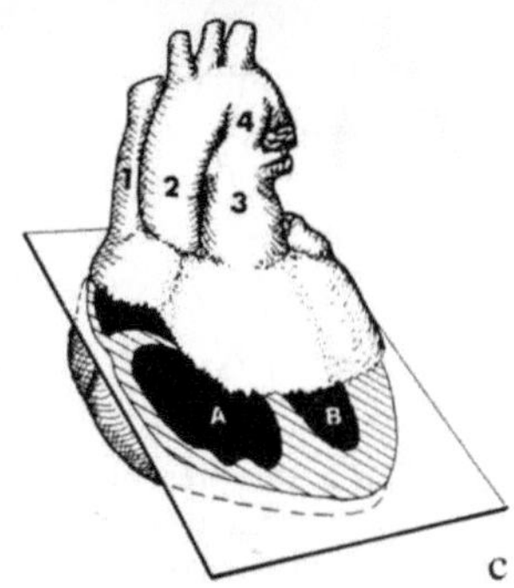

c

Fig. 7.2. a Typical four-chamber view. The fetal spine is at 12 o'clock (for didactic reasons the sonogram was rotated to permit a cranial view of the section)

Fig. 7.2. b Analogous frozen section

Fig. 7.2. c Schematic representation of the four-chamber plane

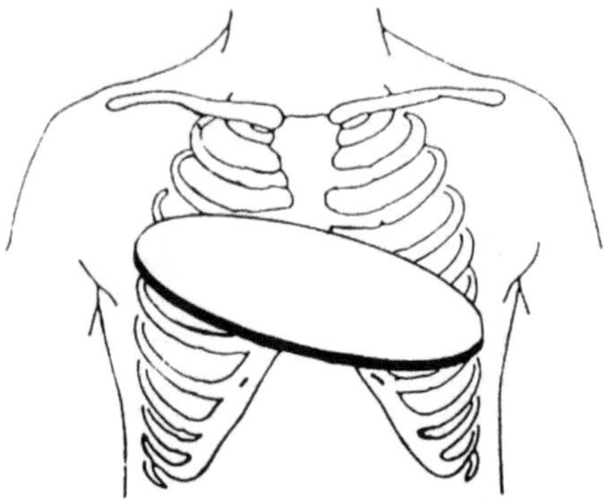

Fig. 7.3. Diagram showing the orientation of the four-chamber plane relative to the thorax

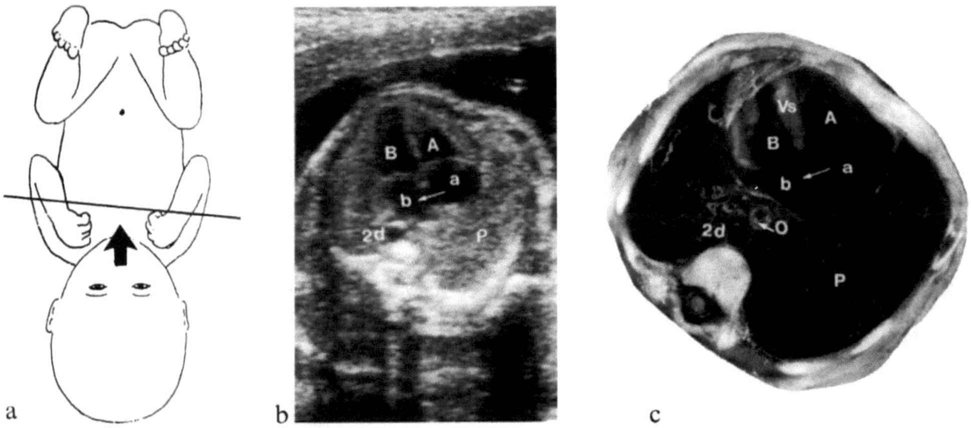

Fig. 7.4. a In this presentation the section is displayed as though the observer were viewing it from above. **b** Four-chamber view displayed as in **a**. **c** Analogous frozen section

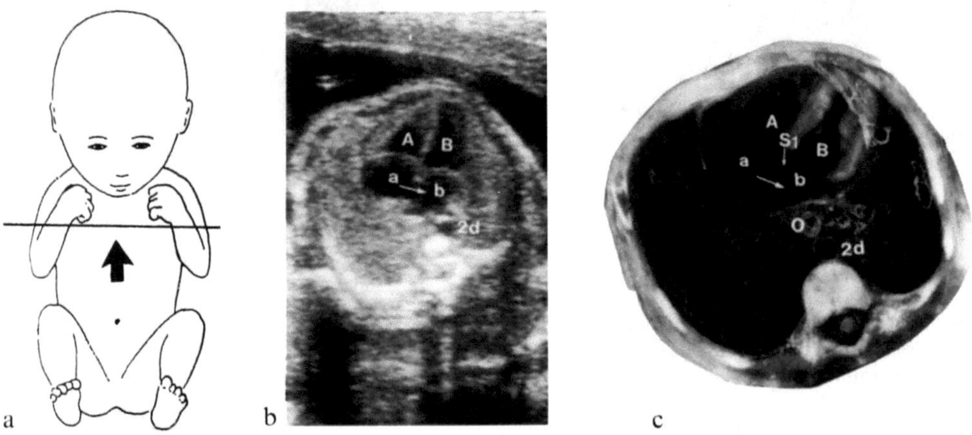

Fig. 7.5. a In this presentation the section is displayed as though the observer were viewing it from below. **b** Four-chamber view displayed as in **a**. The descending aorta appears to the right of the spine. **c** Analogous frozen section

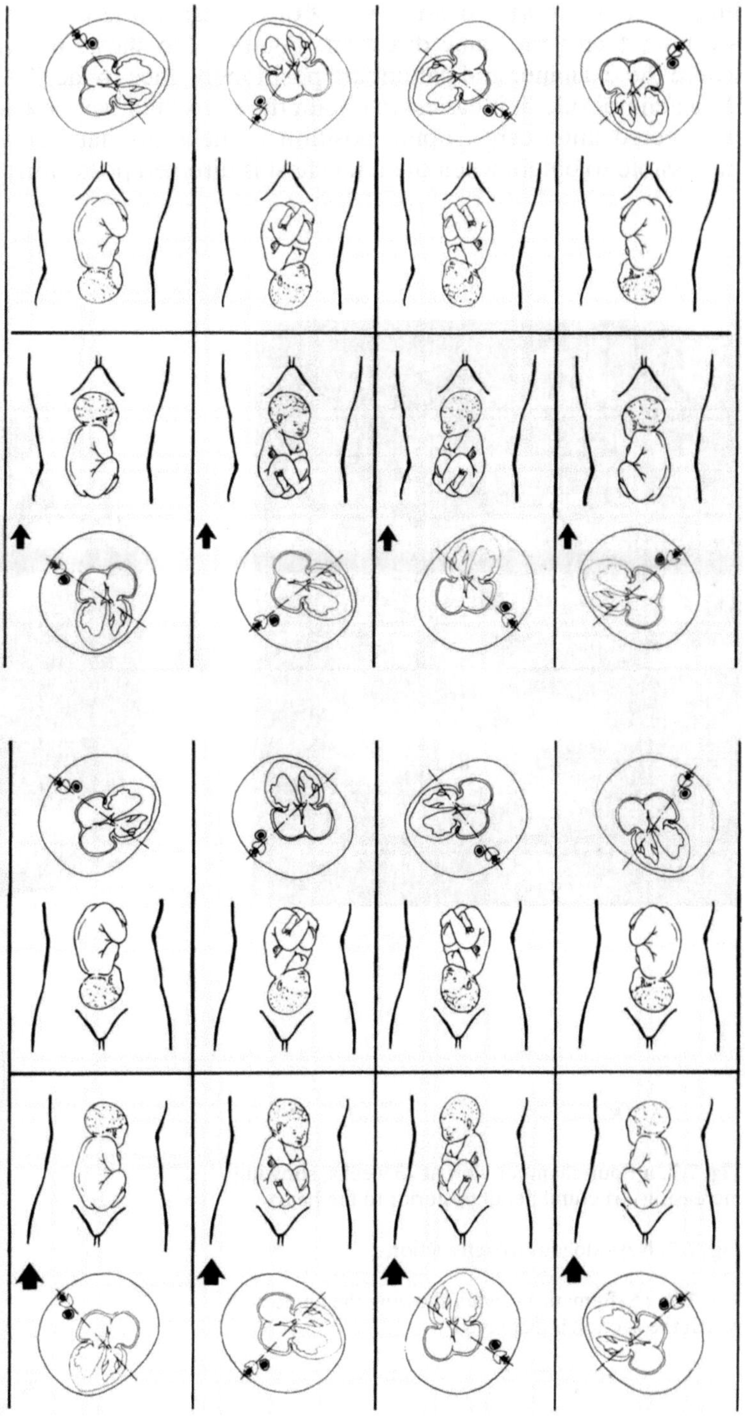

a

b

Fig. 7.6 a, b. Appearance of the four-chamber view for various fetal presentations and positions, **a** with the examiner and the patient facing the same direction toward the monitor screen, **b** with the examiner facing the patient

Figure 7.6 a shows the possible variations for cases where the examiner and patient are both facing the same direction. Figure 7.6 b shows the variations for cases where the examiner and patient are positioned "face to face".
As a general rule, it is easiest to obtain the four-chamber view when the fetal chest is directed anteriorly (supine position). The four-chamber view is difficult or impossible to obtain when the fetal chest is directed posteriorly (prone position).

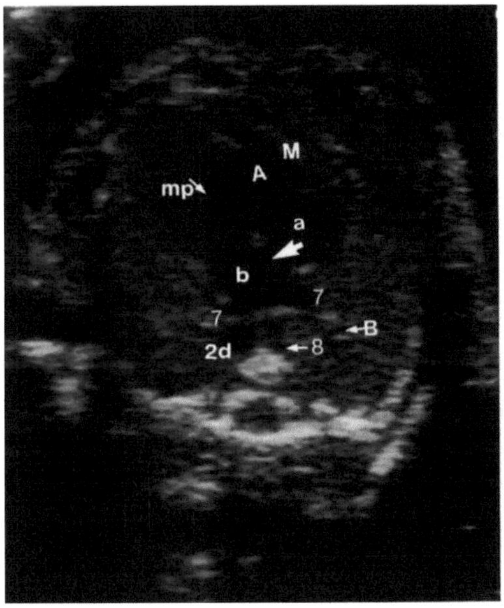

a

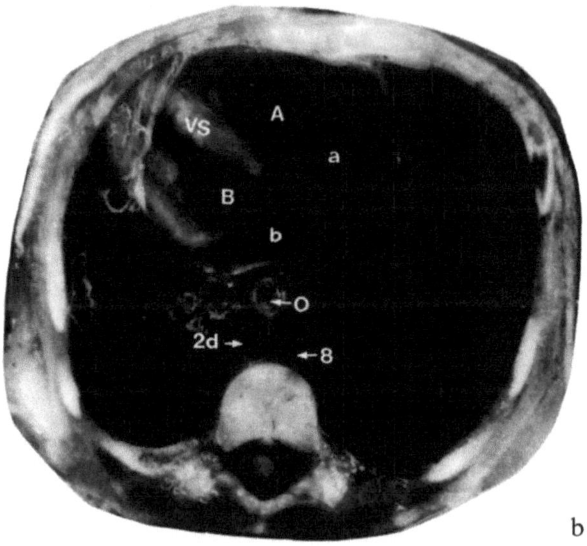

b

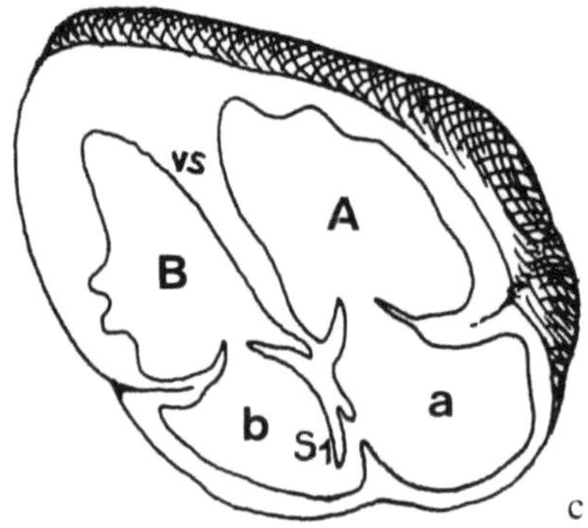

c

Fig. 7.7. a Four-chamber view at 23 weeks, showing increasing structural detail posterior to the heart

Fig. 7.7. b Analogous frozen section

Fig. 7.7. c Schematic drawing showing the angle at which the section is displayed

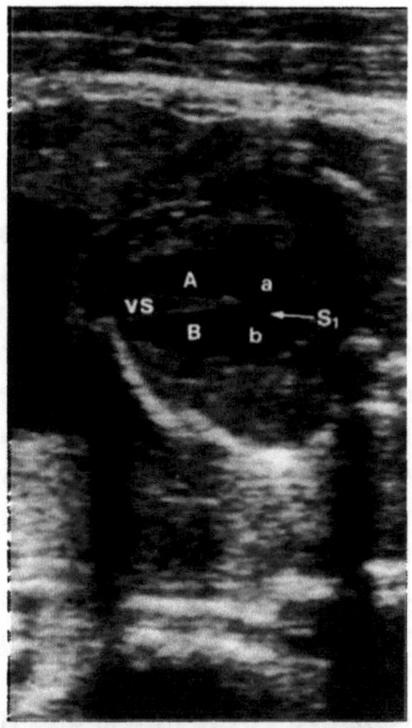

Fig. 7.8. Markedly improved view of the ventricular septum, which in this view is oriented at right angles to the beam

7.3 Four-Chamber View–Vessels and Details

Figure 7.7 a–c shows the typical four-chamber view (section viewed from above) for a supine fetal position. Because the sternum contains only scattered ossification centers prenatally (see Chap. 10), it is not a serious obstacle to scans performed from the anterior aspect. It will be noted that a narrow, hypoechoic rim delineates the myocardium from the lung. Jeanty et al. (1984b) investigated this phenomenon and interpreted it as pericardial fluid. Similarly, we found small pericardial fluid collections in all the thoracic frozen sections that we examined, although it is uncertain whether this was a physiologic phenomenon or a terminal artefact. The ventricular septum is not well defined on this scan, because it is oriented almost parallel to the ultrasound beam. The mitral and tricuspid valves have just closed at the level of the atrioventricular valves, and the foramen ovale is clearly visible between the atria. The motion of the valve of the foramen ovale within the left atrium can be observed directly on real-time scans. The beam intercepts the area of insertion of the pulmonary veins (7) into the left atrium and cuts the descending aorta in front and to the left of the spine. The azygos vein appears to the right of the spine, and bronchial echoes are visible in the hilar region of the lung. The echo structure of the lung appears uniformly granular except for individual bronchi and blood vessels that are traversed by the scan. The esophagus, which presents anterior to the aorta on the frozen section, is not appreciated on the sonogram. Sometimes this view will demonstrate the papillary muscles as small, bright echoes within the ventricles; these should not be mistaken for lesions.

An excellent image of the ventricular septum can be obtained by rotating the fetal spine to the right and the cardiac apex to the left so that the septum is oriented perpendicular to the beam (Fig. 7.8). Leslie et al. (1983) determined the length, width, and thickness of the left and right ventricular wall and ventricular septum in 153 normal aborted fetuses ranging from 15 to 26 weeks' gestation. They then determined the mean ventricular septum/left ventricular wall thickness and ventricular septum/right ventricular wall thickness ratios and found no tendency for those ratios to decrease or increase with advancing gestational age.

7.4 Four-Chamber View–Biometry

The four-chamber view can also be a source of quantitative data for biometry of the fetal heart. Systematic measurements using compound transverse images were published by Levy and Erbsman (1975). Garret and Robinson (1970) and Barret (1979) related the transverse cardiac diameter to the transverse thoracic diameter, determining that the normal ratio was 0.52. Wladimiroff (1981) published the results of cardiac measurements on two-dimensional echocardiograms, and Jeanty et al. (1984a) used similar techniques to determine the normal transverse diameter, longitudinal diameter, and cardiac volume in 695 fetuses examined between 11 and 40 weeks. Their study however remains controversial. DeVore and Platt (1985) warn against the measurement and calculation of cardiac sizes without M-mode techniques, reporting that methodologic errors led to abnormally large cardiac dimensions in approximately 40% of normal fetuses. They recommend that real-time and M-Mode scanning be applied concurrently, and that measurements for the transverse cardiac diameter be made from the M-mode tracing in end-diastole (closure of the mitral and tricuspid valves) at the level of those valves. Figures 7.9 and 7.10 illustrate measurements of the individual cardiac parameters without concomitant M-mode imaging. When cardiac parameters are determined in this way, care should be taken that they are measured by the "outer-outer" dimension, i.e., from epicardium to epicardium. The problems of recording and interpreting M-mode tracings are not within our present scope, and we refer the reader to the pertinent literature (DeVore et al. 1982, 1984, 1985; DeVore and Platt 1985).

7.5 Orientation in Other Planes

Despite frequent claims to the contrary, it has proved very difficult to locate reproducibly other planes of section in the fetal heart, whose nomenclature and topography have been adopted largely from pediatric and adult echocardiography and standardized by Sahn et al. (1978). These planes carry names like "parasternal long axis", "high short axis", "apical four-chamber view", etc. derived from the standard transducer positions for two-dimensional echocardiography in adults (suprasternal, parasternal, apical, subcostal). However, these approaches are not directly accessible in prenatal examinations. Also, in neonatal echocardiography the heart is displayed with the apex on the left and the base on the right. The four-chamber view is always displayed with the apex toward the top of the image. This reference system is feasible only if the examiner sits facing the patient who is being examined. In obstetric sonography, it is customary for the examiner to assume a standard position in relation to the mother (see Examination Setup in Chap. 3). However, the fetus may assume a variety of presentations and positions in utero, and consequently the examiner may encounter a variety of image orientations (see Fig. 7.6 a, b). To follow the conventions of neonatal echocardiography, he would have to find a setup that displayed the fetal heart from the caudal aspect–a measure not feasible in routine prenatal scanning for organizational reasons alone.
To solve these problems, we attempted to establish anatomic reference points that would enable even the less experienced examiner to achieve at least a general ori-

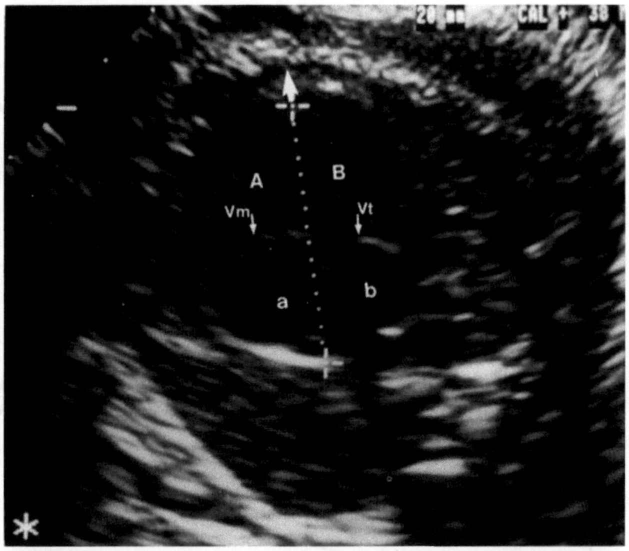

Fig. 7.9. Measurement of the long axis of the heart in the four-chamber view

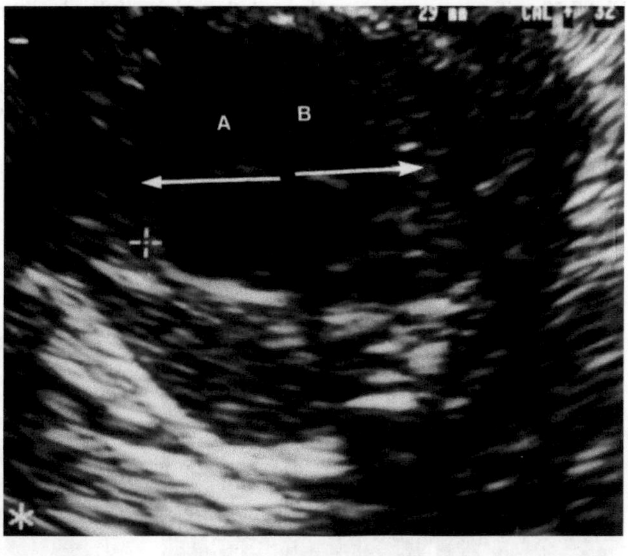

Fig. 7.10. Measurement of the short (transverse) axis of the heart at the level of the atrioventricular valves

entation based on the great vessels of the fetal heart as they appear on sagittal sections. We have found that the transverse section through the region where the three vessels–the superior vena cava, ascending aorta, and pulmonary trunk–enter or leave the base of the heart makes a very good "entry plane" that is easily and reproducibly obtained (Fig. 7.11 a, b). Figure 7.12 a, b illustrates this plane on a sonogram and frozen section. For easier orientation, the section is displayed as though viewed from above. The fetus is in a supine position relative to the examiner. The superior vena cava, ascending aorta, and pulmonary trunk form a series of three ringlike structures that extend obliquely forward from right to left. The scan cuts a portion of the right atrial appendage anterior to the superior vena cava, and the left atrium appears posterior to the three vessels to varying degrees, depending on the

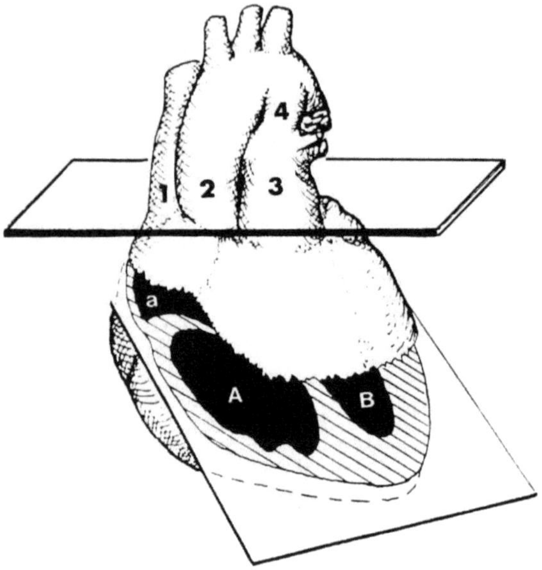

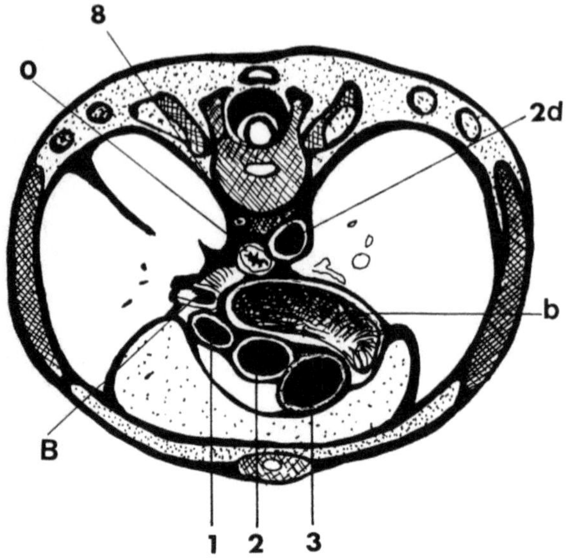

Fig. 7.11. a Transverse section through the origin of the three great vessels (*1* superior vena cava, *2* ascending aorta, *3* pulmonary trunk, *4* ductus arteriosus

Fig. 7.11. b Schematic diagram of the transverse plane of section. The three great vessels appear as an oblique series of ringlike structures. Just posterior to these vessels is the left atrium, and anterior to them is the atrial appendage

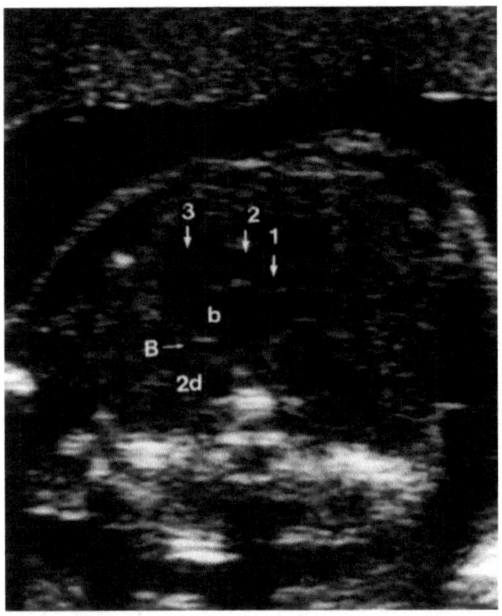

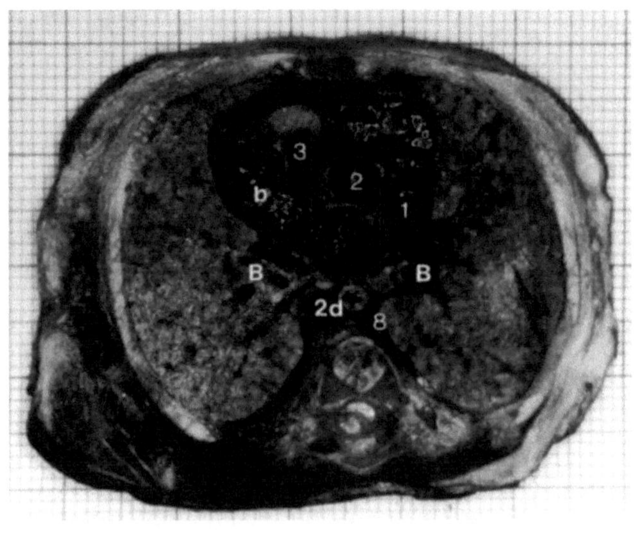

Fig. 7.12. a Sonographic appearance of the transverse section through the base of the heart (supine fetus). The vena cava (*1*), ascending aorta (*2*), and pulmonary trunk (*3*) are lined up obliquely in front of the left atrium (*b*)

Fig. 7.12. b Analogous frozen section from a 21-week fetus

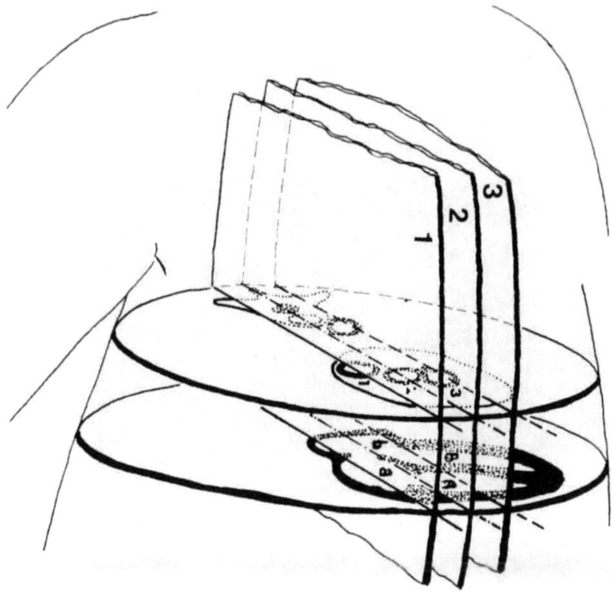

Fig. 7.13. The three sagittal planes of section through the three major vessels

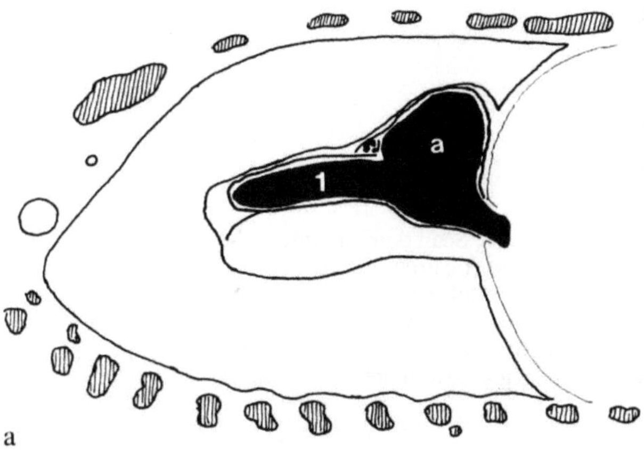

a

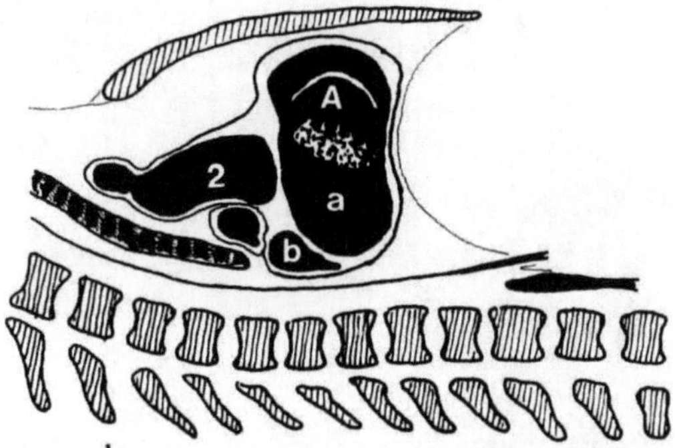

b

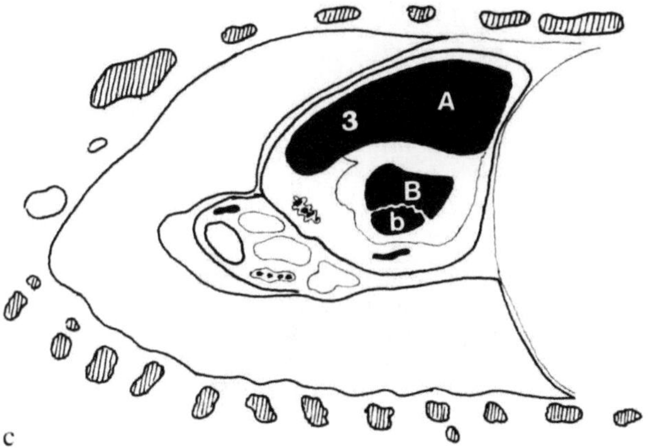

c

Fig. 7.14. a Diagram of the cardiac structures in sagittal plane 1, which displays the superior vena cava, right atrium, and inferior vena cava

Fig. 7.14. b Diagram of the structures in sagittal plane 2, which cuts the right atrium and right ventricle and portions of the left atrium posteriorly

Fig. 7.14. c Diagram of the cardiac structures in sagittal plane 3, which cuts the right ventricular outflow tract; the junction of the left atrium and left ventricle appears posteriorly

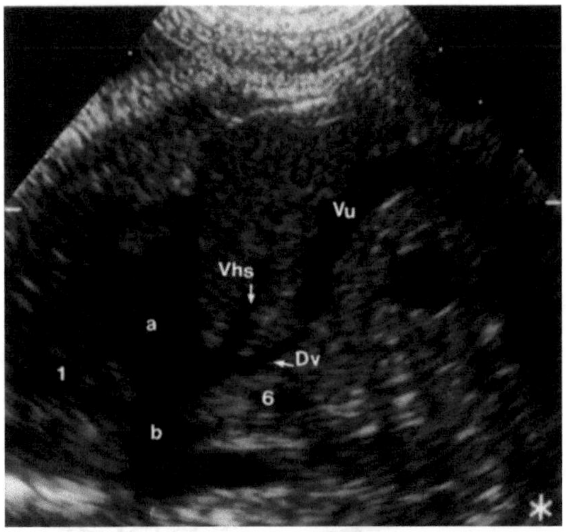

Fig. 7.15. Midsagittal scan in plane 1, demonstrating the superior vena cava, right atrium, inferior vena cava, right hepatic vein, ductus arteriosus, and umbilical vein

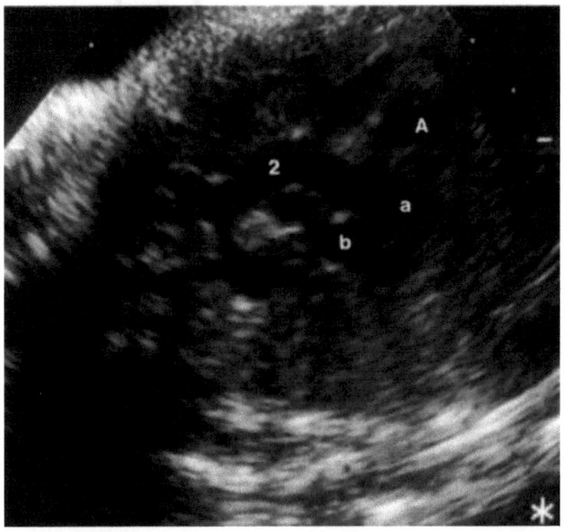

Fig. 7.16. Scan demonstrating the entire aortic arch, including its origin from the left ventricle. As the scan is not tilted but simply rotated about the craniocaudal axis, the right ventricle and right atrium are visualized

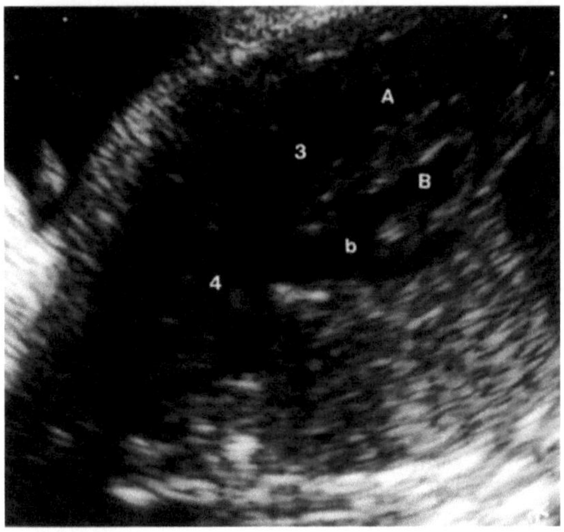

Fig. 7.17. This scan through the right ventricular outflow tract demonstrates the pulmonary trunk as far as the ductus arteriosus. The junction of the left atrium and left ventricle is seen posteriorly, behind the ventricular septum

level of the scan. The descending aorta appears to the left of the spine, the azygos vein appears just in front and to the right of the spine. Between this area and the posterior wall of the left atrium are bright echoes representing the primary bronchi in the hilar region. To train the reader in anatomic orientation, we have drawn three sagittal planes through these vessels (Fig. 7.13), one exactly in the midline and two parasternal, and all angled 90° to the transverse plane of section. We feel that it is important to study these planes, first because they illustrate the complicated topography of cardiac sagittal sections, and second because this orientation forms the basis for understanding the sectional anatomy of the heart.
A sagittal scan through the superior vena cava (Fig. 7.13, plane 1) intercepts the right atrium and the inferior vena cava at its site of entry into the atrium (Fig. 7.14 a). If this scan is tilted even slightly to the left, a portion of the left atrium can be seen through the foramen ovale, posterior to the right atrium (Fig. 7.15).
An exact midsagittal scan cuts a substantial portion of the ascending aorta (Fig. 7.13, plane 2), and part of the trachea may be seen posterior to the aorta. If this scan is angled to the left, the entire aortic arch can be visualized (Fig. 7.16). The right atrium and right ventricle are displayed at this level, with portions of the left atrium appearing posteriorly (Figs. 7.14 b, 7.16). Scans through the pulmonary trunk (Fig. 7.13, plane 3) mainly demonstrate the right ventricular outflow tract and the junction of the left atrium and left ventricle (Fig. 7.14 c). From each of these "standard sections", the scan can be tilted laterally to evaluate the anatomy of the emerging vessels. For example, by tilting the midsagittal scan (Fig. 7.13, plane 2) to the left inferiorly and following the outflow tract into the left ventricle, one can obtain the standard "parasternal long axis" view used in neonatal echocardiography. The mitral valve can be evaluated, and the image plane intercepts the aortic valves in the aortic outflow tract. Part of the right ventricle also is visualized anterior to the interventricular septum. The scan through the pulmonary trunk (Fig. 7.13, plane 3) can be rotated to the right, into the right ventricular outflow tract, to demonstrate the pulmonary trunk as far as the ductus arteriosus (Fig. 7.17).
If suspicious structural patterns are noted on the two-dimensional echocardiogram, it is generally agreed that M-mode scanning and Doppler flow measurements are essential for establishing a diagnosis (Kleinmann et al. 1980, 1982a,b, 1983; Redel and Hansmann 1981; Davis 1982; Sahn 1982; Sahn et al. 1982; Hansmann and Redel 1982; Hansmann et al. 1982; DeVore et al. 1983).

7.6 Malformations

A broad spectrum of developmental anomalies of the fetal heart can be detected and diagnosed by ultrasound (Hansmann et al. 1984, 1985). By obtaining just the four-chamber view, without using any special techniques, it is possible to diagnose the following major anomalies of cardiac development:

1. Hypoplasia of the left ventricle
2. Hypoplasia of the right ventricle
3. Ventricular septal defect (larger than 3 mm)
4. Atrioventricular canal

5. Large atrial septal defect
6. Atresias of the AV valves
7. Premature closure of the foramen ovale
8. Single ventricle
9. Cor biloculare
10. Cardiomyopathies
11. Endocardial fibroelastosis
12. Intra- and extracardiac tumors
13. Dysrhythmias

A detailed description of these anomalies and illustrative sonograms may be found in Hansmann et al. (1985).

8 Abdomen

The topographic anatomy and the separate evaluation of the fetal abdomen during scanning, warrant a chapter on its ultrasound anatomy. The sonoanatomic features of the genitourinary tract are discussed in Chap. 9. Table 8.1 lists the abbreviations that are used to label anatomic structures in the illustrations. As in other body areas, evaluation of the abdomen should begin with a general survey that includes scrutiny of the abdominal outline. This may be done by obtaining an AP sagittal scan through the trunk that is reasonably unobscured by the extremities, or by examining the anterior abdominal surface for defects or protrusions during serial transverse scans.

8.1 Sagittal Sections

The following three sagittal sections are ideal for evaluating the fetal body surface and for general orientation:

8.1.1. Plane 1

This is a right paramedian section through the thorax and trunk (Fig. 8.1 a–c). The lung and liver are distinguished from each other by their contrasting echo densities. Kossoff (1981) reports that this reflectivity changes with gestational age. He found that lung reflectivity was less than that of the liver in the immature fetus, was equal to that of the liver from 34 to 37 weeks, and exceeded that of the liver thereafter. We personally were unable to reproduce this finding in our examinations, and to us the lung always appeared denser than the liver, regardless of gestational age.

Table 8.1. List of abbreviations used for intra-abdominal organs and blood vessels

Diaphragm	D	Descending aorta	2d
Liver	H	Inferior vena cava	6
Spleen	L	Right hepatic veins	Vhd
Kidney	R	Left hepatic veins	Vhs
Adrenal gland	Gs	Umbilical vein	Vu
Pancreas	Pr	Ductus venosus	Dv
Stomach	V (M)	Portal vein	Vp
Gallbladder	Vf (GB)	Portal sinus (left portal vein)	Vps
		Right portal vein	Vpd

Benson et al. (1983) also believe that fetal lung maturity can be assessed from changes in echogenicity, but they used very sophisticated instruments in their studies (radiofrequency analysis), and the results cannot be applied to the images produced in routine scanning.
The contrasting echo patterns of the liver and lung provide an indirect delineation of the fetal diaphragm (on both sagittal and frontal scans; see Fig. 8.13 a). Aspects of the ultrasound diagnosis of diaphragmatic defects have been discussed by Jeanty and Romero (1984) and by Hansmann et al. (1985). This diagnosis includes not only visualization of the defect itself but also of associated anatomic displacements. Often this involves the herniation of stomach and bowel into the left side of the fetal thorax with concomitant right-sided displacement of the heart. If the paramedian sagittal scan is performed close enough to the right side of the body, the kidney can be visualized behind and caudal to the liver (Fig. 8.1 a–c).

8.1.2 Plane 2

This parasagittal section is obtained by moving the transducer past the midline toward the fetal left side (Fig. 8.2 a–c). In the thoracic region the scan may show only lung, or it may intercept a small portion of the left ventricle, depending on the distance of the image plane from the midline. Caudal to the diaphragm, the stomach can be seen below the left lobe of the liver. As the transducer is moved farther toward the fetal left side, the left kidney will come into view. The basic guidelines for the anatomic and functional evaluation of the fetal stomach by ultrasound were established by Wladimiroff et al. (1980), Vandenberghe and DeWolf (1980), and Bowie and Clair (1982). Generally the stomach can be visualized from 12 weeks on. Wladimiroff et al. (1980) investigated fetal stomach filling and emptying times. They found that the stomach filling time generally was less than 45 min, while emptying times ranged from a few minutes to 30 min, taking less than 5 min in the majority of cases. Vandenberghe et al. (1980) used the formula for the volume of a cylinder ($V = 0.785$ abc) to calculate the fetal stomach volume. They found mean stomach volumes of 1 ml at 20 weeks, 2 ml at 25 weeks, 5 ml at 30 weeks, and 8 ml at 35 weeks. The upper limit of normal in the third trimester was 10 ml at 30 weeks and 22 ml at 35 weeks. They observed abnormally large stomach volumes in fetuses with duodenal atresia and in one case calculated a volume of 22 ml before 30 weeks. However, the significance of this finding is qualified by the observations of Bowie and Clair (1982), who showed that the stomach volume can be affected by fetal vomiting occurring in association with duodenal atresia and other high gastrointestinal obstructions.

8.1.3 Plane 3

Abdominal plane 3 is precisely midsagittal (Fig. 8.3 a–c). In the thorax it cuts the heart and related structures, and slightly paramedian sections will travers the junction of the inferior vena cava and right atrium (see Chap. 7). Scans in this plane will often demonstrate the umbilical vein from its site of entry into the abdomen to its junction with the ductus venosus. The inferior vena cava appears as a hypoechoic band like structure passing obliquely upward and forward. By moving the beam plane slightly, the ductus venosus can be demonstrated as a direct communication between the end of the umbilical vein and the left hepatic vein.

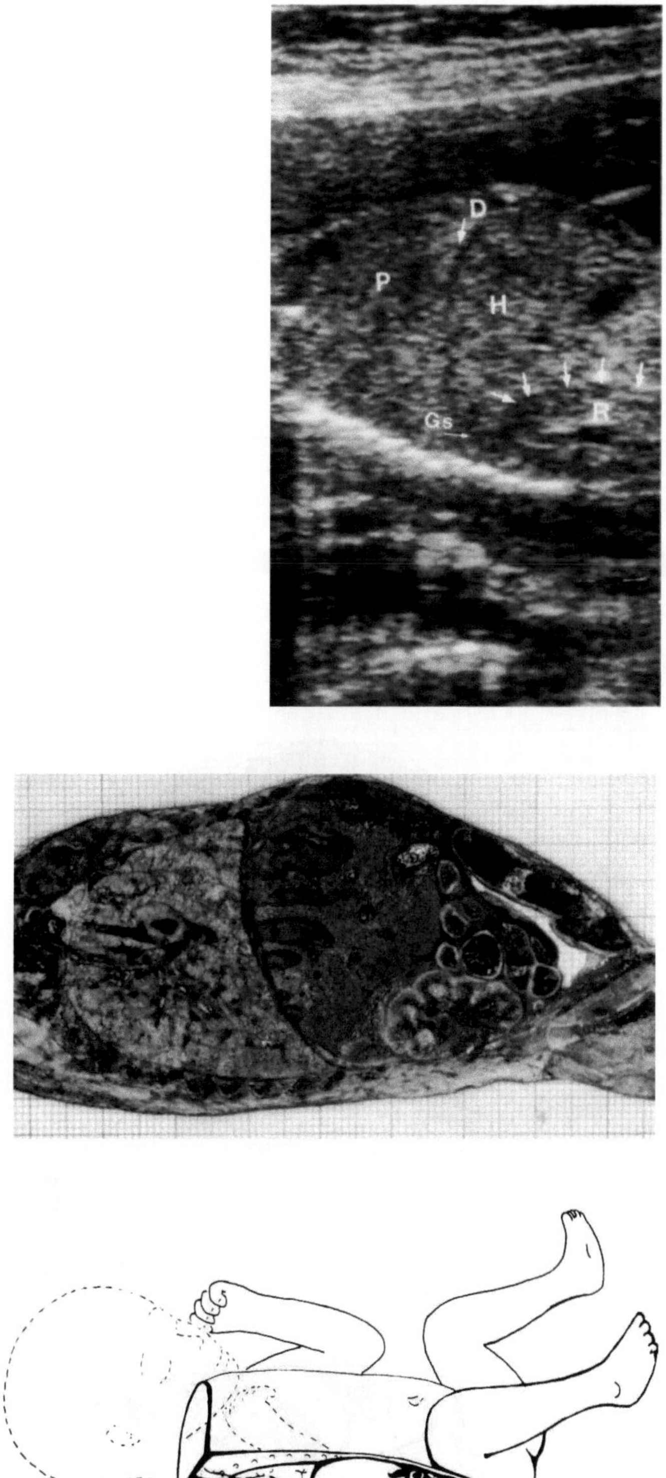

Fig. 8.1 a–c. Sagittal section of the abdomen in plane 1. **a** Scan of a 22-week fetus. **b** Frozen section at 22 weeks. **c** Schematic drawing

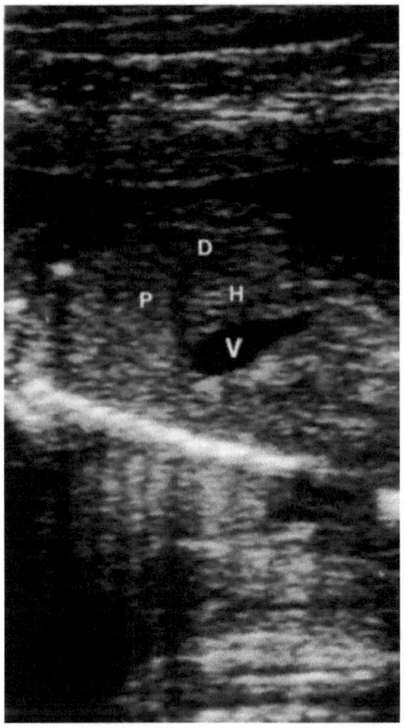

Fig. 8.2. a Sagittal scan plane 2 in a 20-week fetus

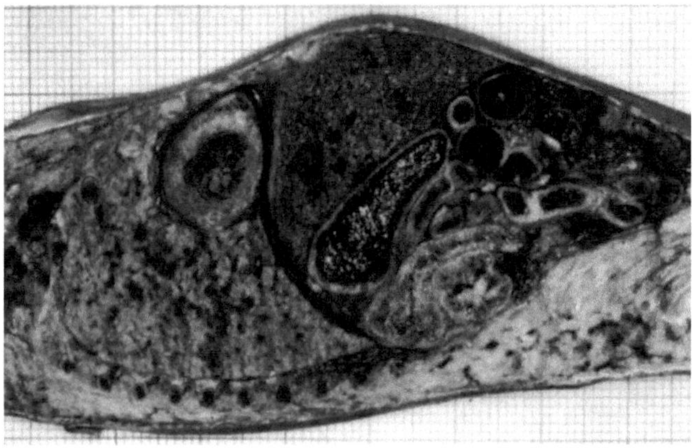

Fig. 8.2. b Analogous frozen section at 21 weeks. The section cuts part of the left ventricle, which is not visible on the sonogram

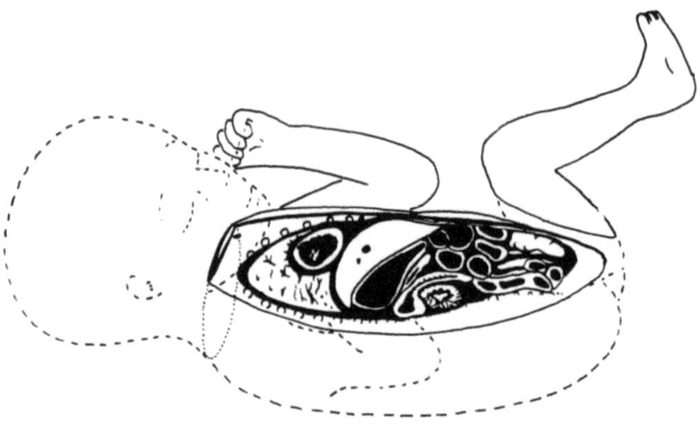

Fig. 8.2. c Schematic drawing of the plane of section

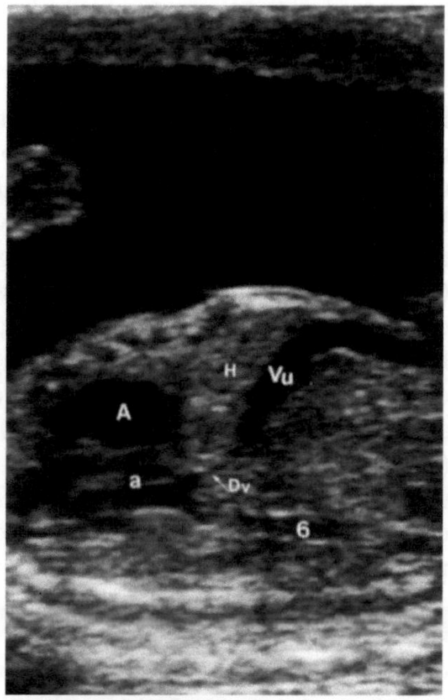

Fig. 8.3. a Midsagittal scan through the thorax and abdomen of a 23-week fetus

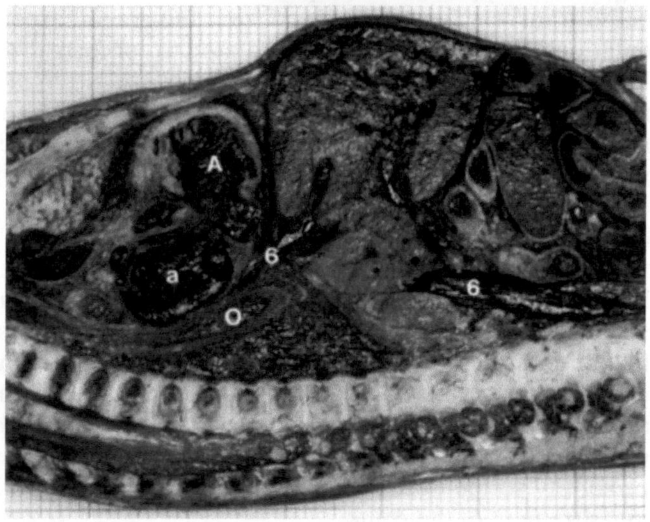

Fig. 8.3. b Analogous frozen section at 22 weeks

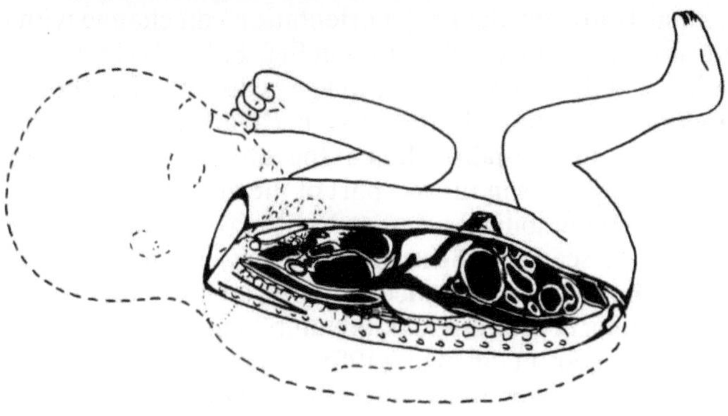

Fig. 8.3. c Schematic drawing

8.2 Venous Vessels

Knowledge of the vascular anatomy of the fetal liver is important as a basis for orientation, especially in connection with trunk biometry. Experimental studies in corrosion preparations of the fetal liver by Kugener and Hansmann (1976) and Morin and Winsberg (1978) along with sonoanatomic studies of the intra-abdominal vessels by Morin and Winsberg (1978), Bernaschek et al. (1980), Chinn et al. (1982), and Jeanty et al. (1984) have laid the essential groundwork for accurate orientation in this region. To interpret correctly the sonoanatomic features of the fetal intra-abdominal venous architecture, it is necessary to know the basic anatomy of the umbilical vein, portal vein, and inferior vena cava. Figure 8.4 shows a schematic drawing of the principal vessels of the fetal liver viewed from anterior (Moore 1977), and Fig. 8.5 shows a posterior view of the fetal liver as it appears in most anatomic textbooks. Figure 8.6 shows a diagram of the portal venous branches viewed from anterior. Part of the blood from the umbilical vein enters the portal sinus (left portal vein). The portal sinus is the only venous vessel that presents much of its length on a true transverse section. Morin and Winsberg (1978) stressed the importance of the portal sinus for accurate trunk measurements based on their studies in vascular corrosion preparations. Figure 8.7 a–c shows this region as it appears sonographically and in anatomic section. The section is oriented transversely and is viewed from the cranial aspect (Fig. 8.7 c). (To aid orientation, the sonogram has been rotated to conform to the anatomic presentation.) Major sonographic landmarks at this level are the spine, the vena cava coursing in front and to the left of the spine, and the abdominal aorta coursing slightly to the right. The stomach appears in the left half of the abdomen, and the scan exactly traverses the confluence of the umbilical vein and portal sinus. The anatomic section is at a slightly lower level than the sonogram, and the umbilical vein has not yet entered the portal sinus. The adrenal glands appear on either side of the spine on the anatomic section, and part of the pancreas is visible between the stomach and left adrenal. We have rarely been able to identify the pancreas directly in our ultrasound examinations. Jeanty et al. (1984), in their studies of fetal vascular anatomy, have noted the possibility of identifying the pancreas using the splenic vein as a guide. If the scan is moved to a slightly higher level and viewed from the cranial aspect, the umbilical vein is seen to move anteriorly while the kidneys appear on either side of the spine, and often the gallbladder appears on the right side in proximity to the liver (Fig. 8.8 a, b). This scan is too low to permit an accurate measurement of the trunk.

To illustrate how the image orientation can change with the fetal presentation, we show a similar plane of section in Fig. 8.9 a, b viewed from the caudal aspect. The spine is at 5 o'clock, and the image is displayed as though viewed from below. The abdominal aorta and inferior vena cava are cut transversely in front of the spine. The scan cuts the gallbladder at the inferior border of the right lobe of the liver, the stomach appears in the left part of the abdomen, and the kidneys present paravertebral. The umbilical vein lies just below the anterior body surface. By tilting the transverse scan about the right-left axis, it is possible to image the entire length of the umbilical vein, provided it follows a reasonably straight course (Fig. 8.10 a, b). [Variations in the course of the umbilical vein with gestational age have been described by Hansmann et al. (1985).] Again the image is displayed as though the section were being viewed from below. The umbilical vein courses posteriorly from

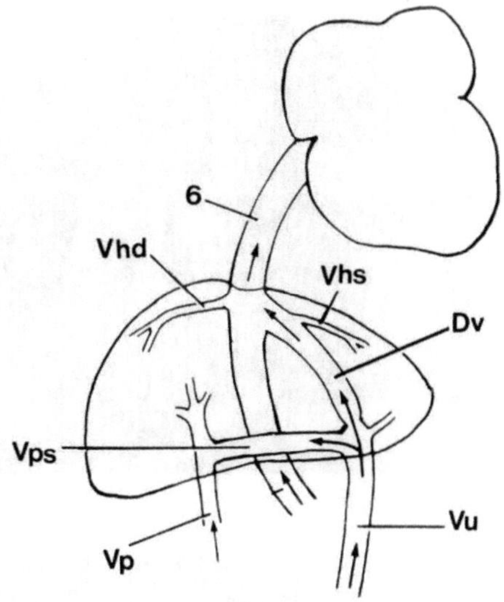

Fig. 8.4. Schematic diagram of the venous anatomy of the fetal liver, anterior aspect (after Moore 1977)

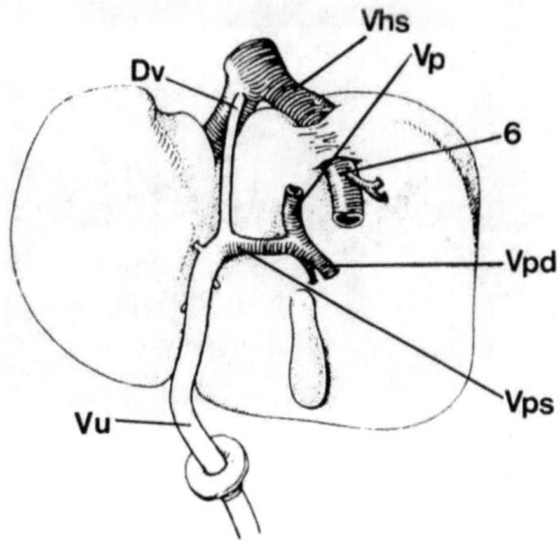

Fig. 8.5. Posterior view of the area of the umbilical vein-portal sinus confluence

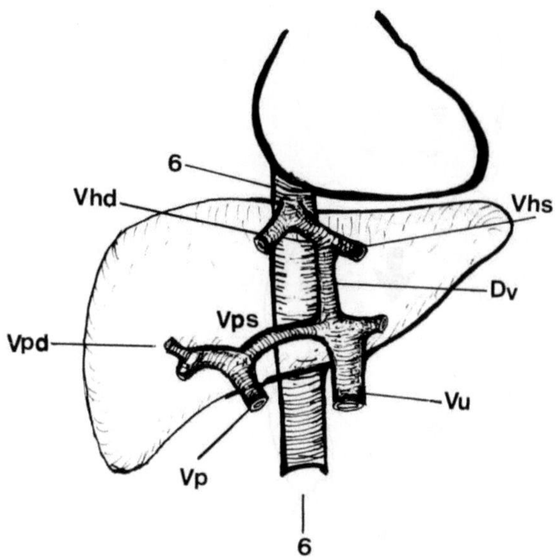

Fig. 8.6. Diagram showing the relationship of the major vessels to the ductus venosus

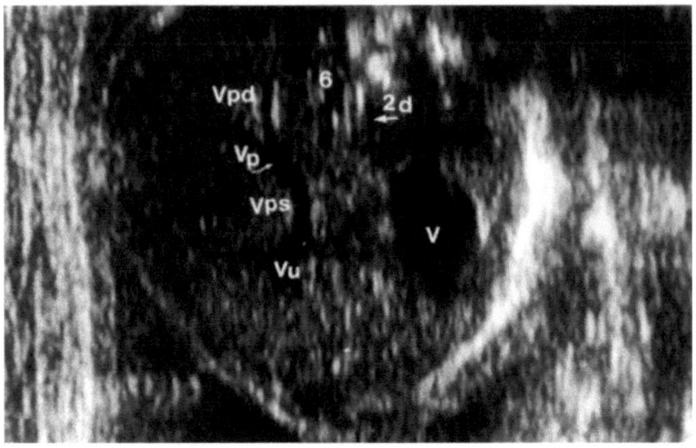

Fig. 8.7. a Transverse scan through the fetal abdomen in the plane for trunk biometry

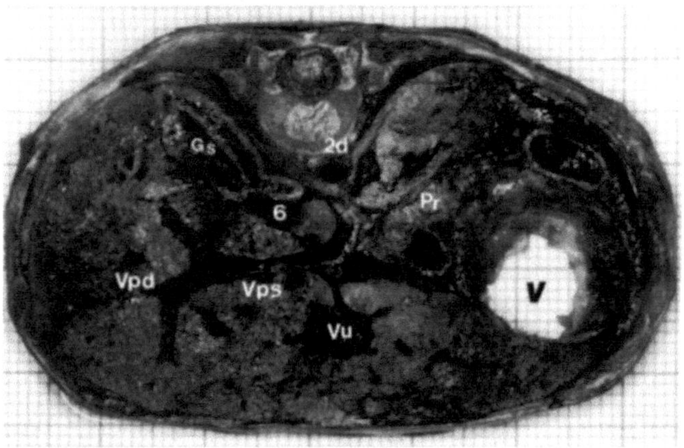

Fig. 8.7. b Frozen section at an analogous level in a 20-week fetus

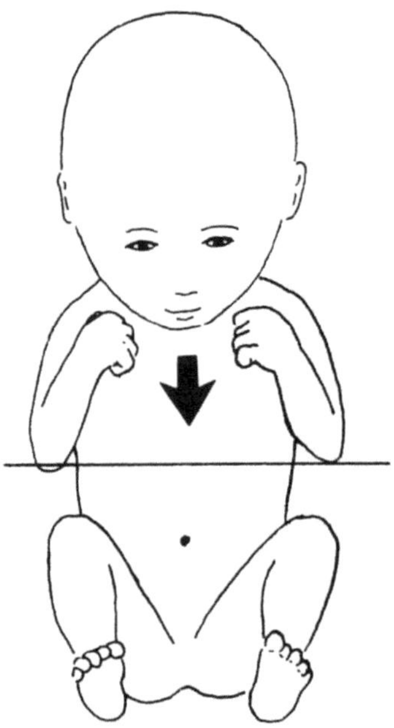

Fig. 8.7. c Diagram showing how the scan is displayed to the viewer

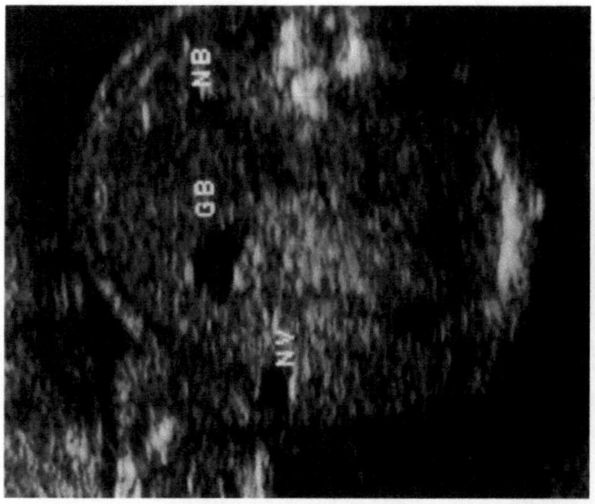

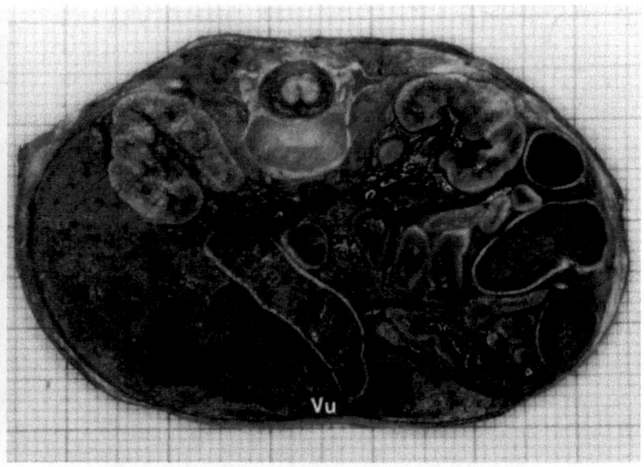

Fig. 8.8. a Transverse scan through the fetal abdomen at 22 weeks (rotated 90° for easier orientation). The scan is caudal to the reference plane for trunk biometry and traverses the kidney, liver, gallbladder (*GB*), and umbilical vein insertion (*VU*)

Fig. 8.8. b Analogous frozen section at 21 weeks

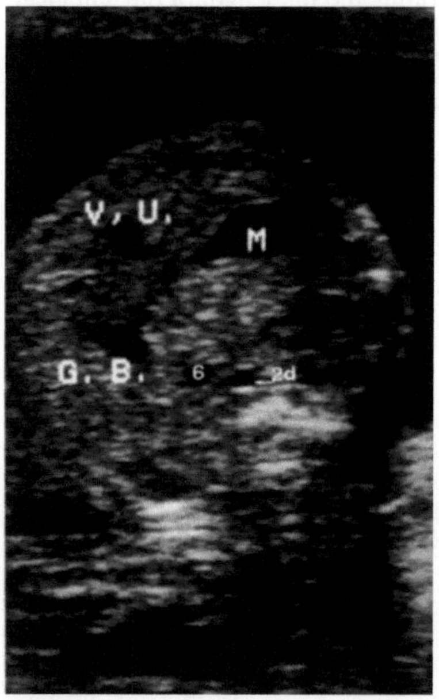

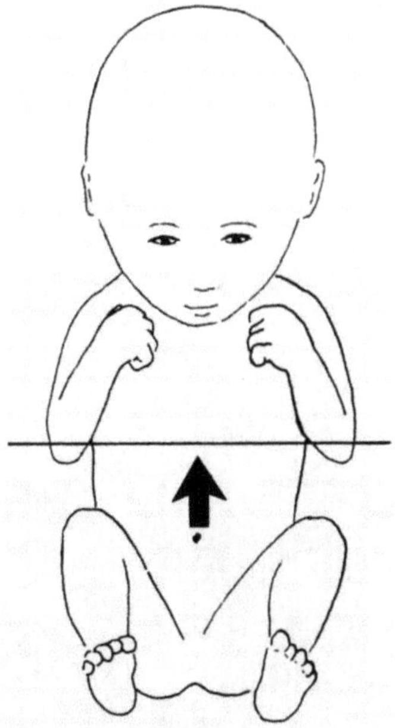

Fig. 8.9. a Transverse scan through a fetal abdomen caudal to the plane for trunk measurement. The fetus is supine (spine at 5 o'clock), and the section is viewed from below, as shown schematically in **b**. The umbilical vein, stomach (*M*), and gallbladder (*GB*) are demonstrated

Fig. 8.9. b Diagram showing how the scan is displayed to the viewer

posteriorly from the anterior body surface and drains into the portal sinus. The gallbladder appears anterior to it. A short segment of the ductus venosus is visible owing to the obliquity of the image plane. If the scan is tilted further so that it approaches a frontal orientation, it may display portions of the hepatic veins in the area where they enter the inferior vena cava just below the heart (Fig. 8.11).

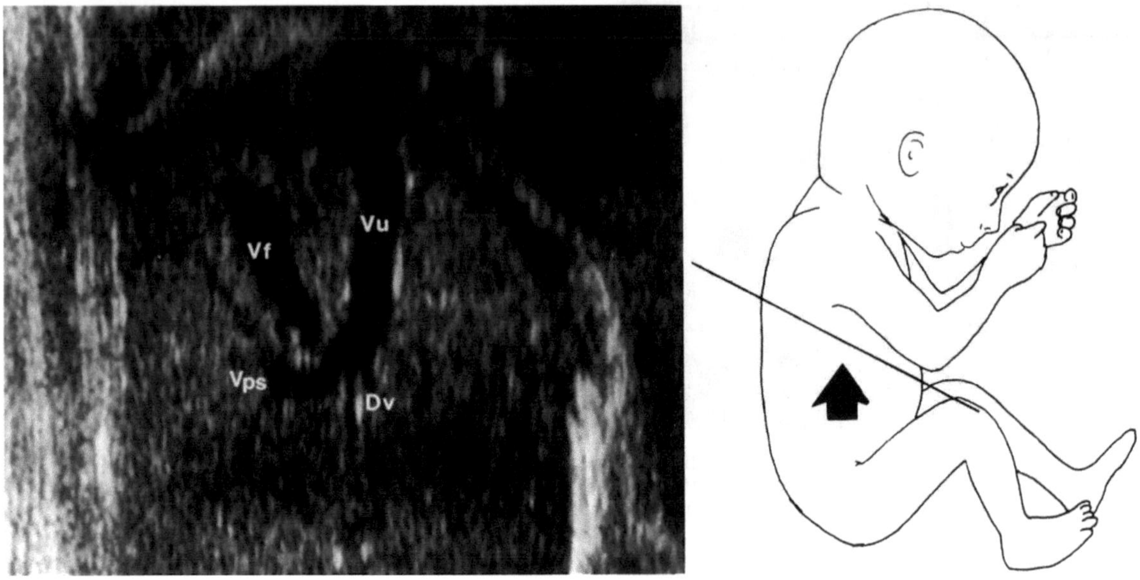

Fig. 8.10. a Oblique scan through a fetal abdomen demonstrating the full length of the umbilical vein and the origin of the ductus venosus. The gallbladder (*Vf*) is seen to the left of the umbilical vein. The oblique section is displayed as if viewed from below, as illustrated in **b**

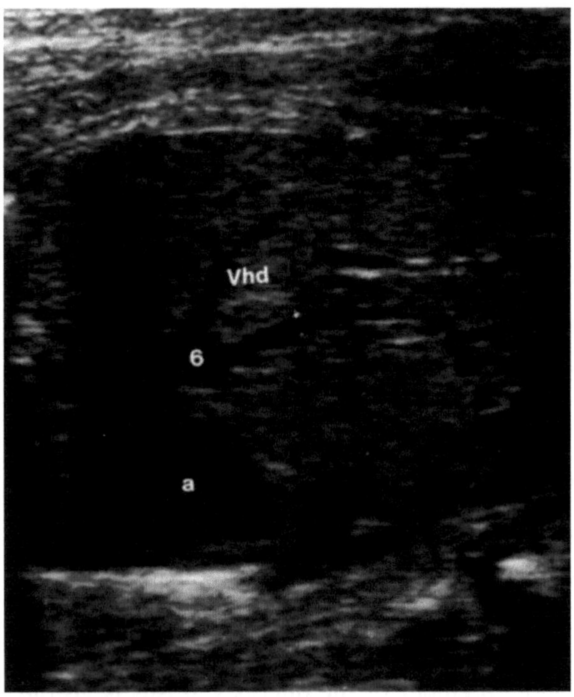

Fig. 8.11. Oblique, almost frontal scan through a fetal abdomen displaying the hepatic veins just before their entry into the right atrium

8.3 Reference Plane for Trunk Biometry

Measurement of the fetal trunk should be a standard part of every examination, even routine examination. Different authors prefer different trunk parameters or their combinations. The first measurements of the fetal trunk were performed by Thompson et al. (1965), and since then many authors have stressed the importance of trunk measurement in the assessment of gestational age, weight estimation, and the detection of intrauterine growth retardation (Garret and Robinson 1971; Bayer et al. 1972; Holländer 1972, 1975, 1984; Hansmann and Voigt 1973; Schlensker and Decker 1973; Schillinger et al. 1975; Levi and Erbsman 1975; Campbell and Wilkin 1975; Hansmann 1975; Higginbottom et al. 1975). We favor measurement of the transverse trunk diameter for routine examinations. The reference planes are easily identified, and the results are reproducible (even among different examiners) when proper attention is given to anatomic landmarks and technical detail. Our procedure for measuring this parameter is as follows:

1. A longitudinal scan is performed to establish the fetal longitudinal axis.
2. The transducer is rotated 90° from this axis on the maternal abdomen, and the heart is imaged without obtaining a four-chamber view.
3. The transducer is moved caudally from the level of the heart until the umbilical vein appears.
4. Parallel cuts are made along the course of the umbilical vein to locate the point where the typical oval outline of the vein disappears or is just visible.
5. The transducer pressure on the maternal and thus on the fetal abdomen is kept to a minimum to avoid changes of outline due to compression.
6. The examiner again confirms that a true transverse, symmetrical section has been obtained before the scan image is frozen for measurement.

Several confirmatory criteria are applied: 1) The section should have an approximately circular outline, 2) the rib echoes should extend equal distances on both sides, and 3) three ossification centers should be visible in the spine. The transverse trunk diameter is measured on the frozen image after positioning the cursor on the outer contours of the section (Fig. 8.12).

Significant measurement errors do not result from tilting of the scan on the right-left axis, and indeed the distance between the points of measurement would not change even if the scan were rotated to a frontal orientation (Fig. 8.13 a, b). However, if the scan is tilted laterally (rotated on the AP axis), the measurement error will increase with the degree of the tilt. Figure 8.14 shows the trunk diameter as it appears on a laterally tilted scan. Applying the criteria stated above, we find several indications of faulty plane selection: 1) the overall shape of the section is elliptical, 2) the characteristic triangular cluster of vertebral ossification centers is absent, and 3) the rib echoes have unequal appearance and length.

In experimental measurements on frozen fetuses, we found that rotating the plane of measurement 30° on the AP axis had the effect of increasing the apparent transverse trunk diameter from 75 mm to 83 mm, producing an error of approximately 11% (Fig. 8.15 a, b). Applying this result to Hansmann's tables of normal values (Hansmann et al. 1985), we find that the correctly measured transverse trunk diameter (75 mm) corresponds to about 29 weeks' gestation, while the diameter

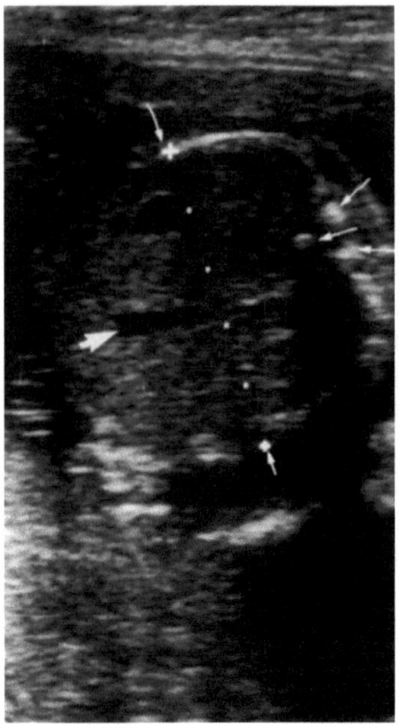

Fig. 8.12. Trunk biometry in a 19-week fetus. The arrows mark the check points for establishing the correct transverse plane

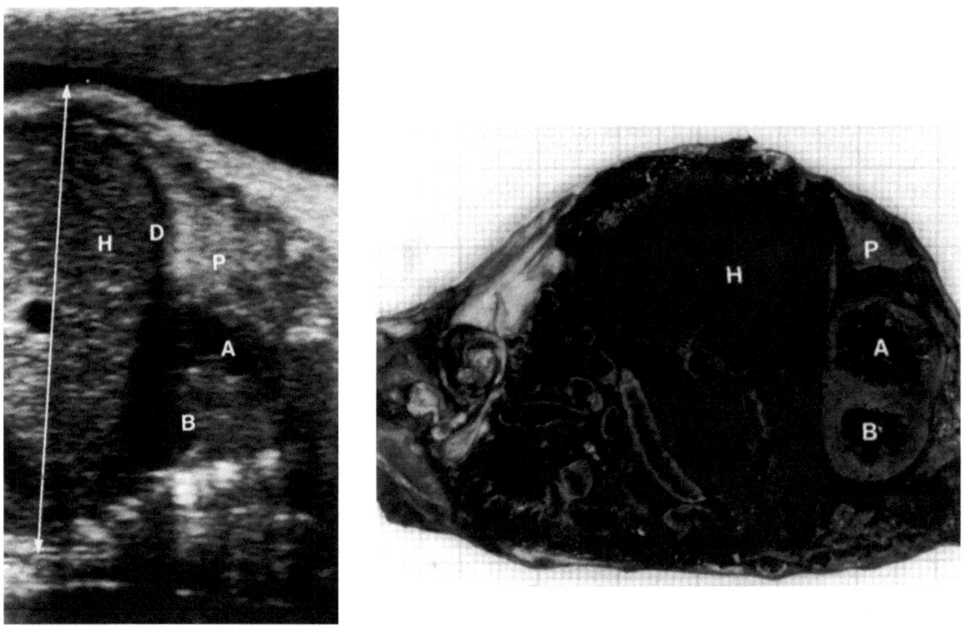

Fig. 8.13. a Fetal trunk biometry on a frontal scan. The diaphragm is visible, and the umbilical vein appears in the center of the liver as a reference point for the correct plane

Fig. 8.13. b Analogous frozen section at 22 weeks

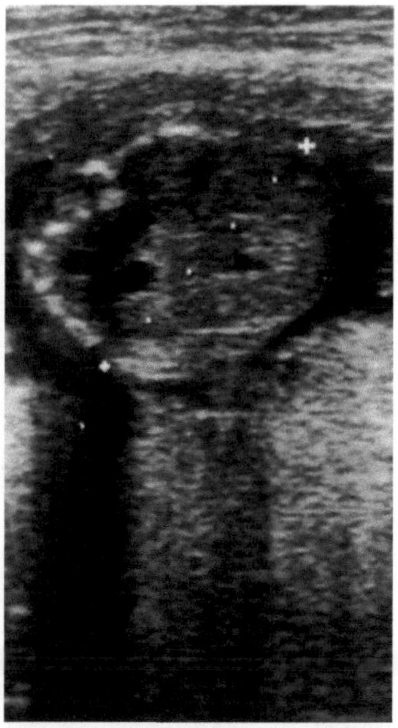

Fig. 8.14. Incorrect measurement of the transverse trunk diameter

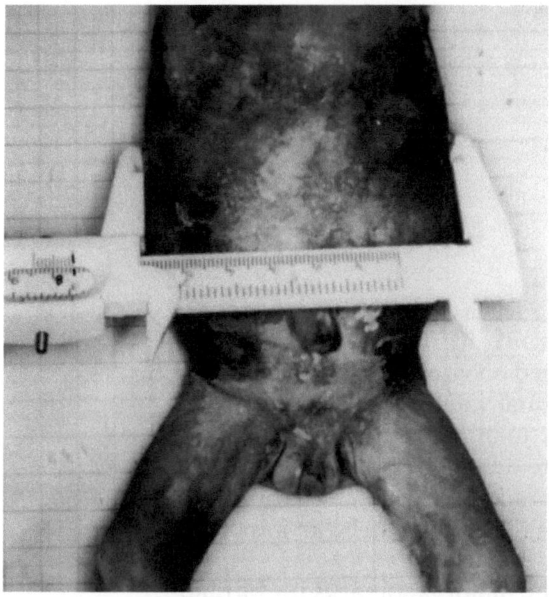

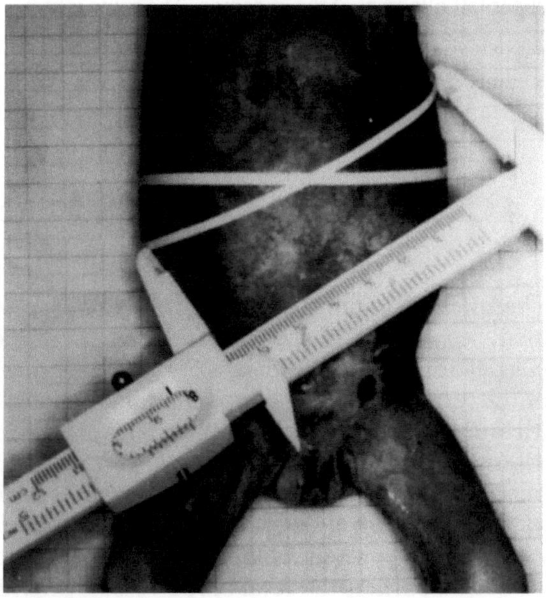

Fig. 8.15. a Correct measurement indicates a transverse trunk diameter of 75 mm

Fig. 8.15. b The plane of measurement has been rotated 30° about the AP axis, yielding a transverse diameter of 83 mm

measured at a 30° tilt (83 mm) corresponds to 32 weeks. When the trunk diameter is measured in the AP axis, analogous errors result from tilting the scan about the horizontal axis (see Chap. 3).

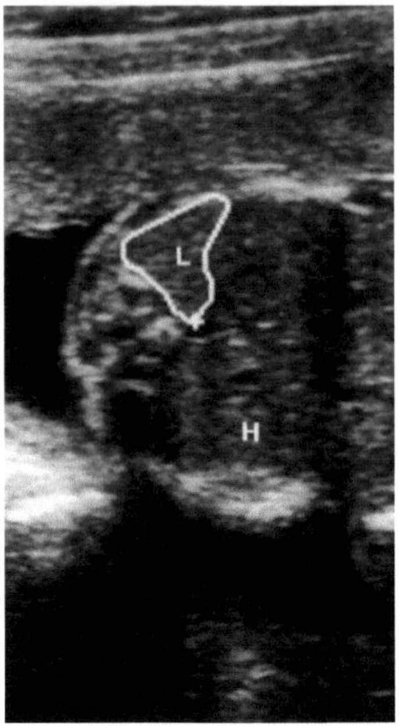

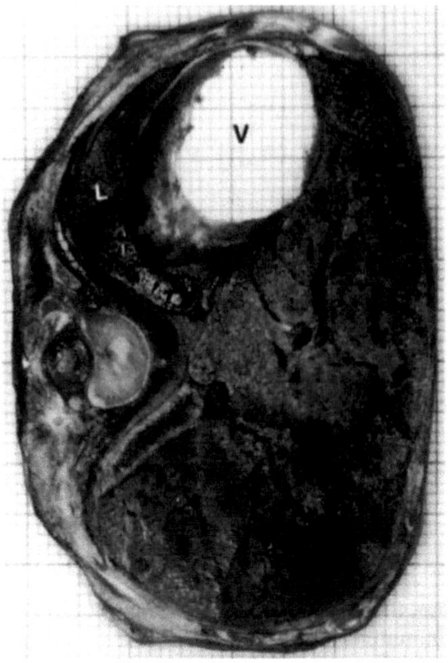

Fig. 8.16. a Transverse scan through a fetal trunk at 23 weeks, demonstrating the spleen (*L*) in the upper left quadrant of the abdomen. The spine is at 9 o'clock

Fig. 8.16. b Analogous frozen section, which includes the cavity of the stomach (*V*). The stomach is empty on the sonogram

8.4 Other Intra-Abdominal Structures

The spleen also may be demonstrated on transverse scans through the upper abdomen, especially if the left side of the fetal trunk is facing the transducer (Fig. 8.16 a, b). Schmidt et al. (1985) have published on visualizing and measuring of the fetal spleen, and corresponding tables of normal values have been published (Schmidt et al. 1985; Hansmann et al. 1986). Certainly, spleen biometry is not essential for routine examinations. But according to Schmidt et al. (1985) it has an important role in evaluating the severity of Rh incompatibility and in the diagnosis of syndromes involving the spleen. Enlargement of the spleen can be an important diagnostic sign when fetal infection is suspected.

As stated earlier, the fetal blood vessels provide the major landmarks for intraabdominal sonography. The importance of the abdominal aorta as a guide to visualizing the kidneys on frontal scans is discussed in Chap. 9. The frontal scan in Fig. 8.17 displays the entire abdominal aorta with its division into the right and left common iliac arteries.

The key aspect of a basic examination is the correct identification of fluid-filled intraabdominal structures (stomach, bladder). The finding of a fluid collection at a site not corresponding to normal anatomy justifies referral for detailed assessment. Whenever the examiner detects an apparent cystic lesion, he should examine the kidneys as a means of relating the lesion to the gastrointestinal or genitourinary

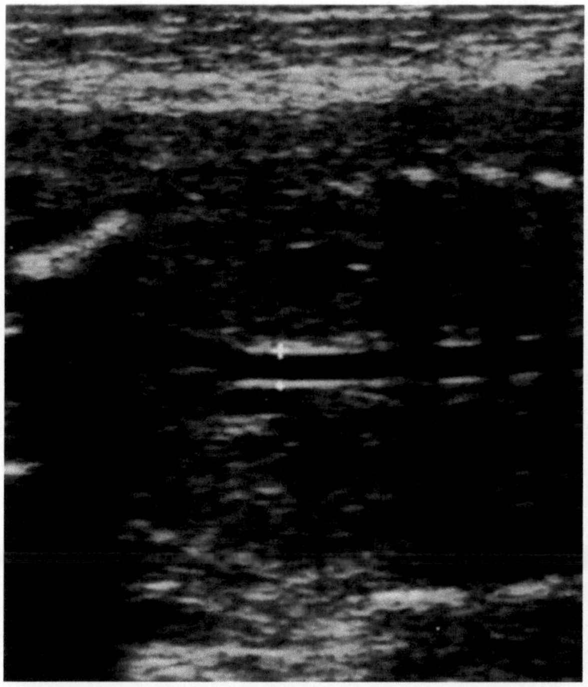

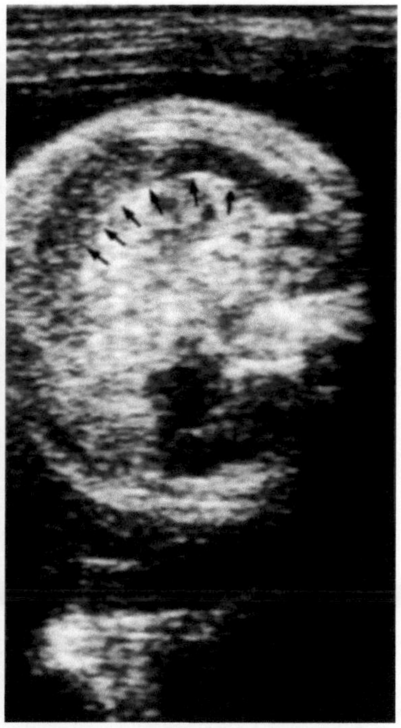

Fig. 8.17. (left) Frontal scan through a fetal abdomen at 23 weeks, showing the abdominal aorta and its division into the right and left common iliac arteries

Fig. 8.18. (right) Transverse scan through a fetal trunk at 39 weeks. The fluid-filled transverse colon *(arrows)* courses below the anterior abdominal wall

tract. If the kidneys appear normal, the observed cyst(s) must involve the gastrointestinal or genital tract (mesenteric cyst, choledochal cyst, splenic cyst, ovarian cyst). The anatomic structures of the intestinal tract can rarely be demonstrated in detail before the third trimester. Usually the colon is the first structure to be identified by virtue of its size and haustration (Fig. 8.18). Prolonged observation in the third trimester may disclose sequences of peristaltic motion of the stomach and bowel. This is important, because temporarily distended bowel loops can mimic a cystic lesion, and if the examination is continued for a sufficient period, the "mass lesion" often will be cleared by peristalsis.

If the fetus is lying prone with the arms and legs flexed, it can be extremely difficult to evaluate the abdominal contents by ultrasound. But because accurate trunk measurement is a crucial part of the examination, and congenital anomalies of the genitourinary and gastrointestinal tract are fairly common, an attempt should be made to gain access to this region. Initially, the patient should be positioned on her side. If this is unsuccessful, the examiner can try percussing the maternal abdomen to provoke fetal movements in the hope of creating satisfactory windows to the structures of interest. If this still does not yield the desired result, the examinion should be repeated at a later time; this is imperative in view of the above statements.

9 Genitourinary Tract

Evaluation of the fetal genitourinary tract is of major importance due to the prevalence of anomalies in that area. In our study population, genitourinary malformations were very common and second only to central nervous system defects. The overall incidence of urinary tract malformations found in pediatric autopsies ranges between 5% and 7.8% (Zschock et al. 1968; Porter 1978).

The importance of the prenatal ultrasound evaluation of this region is demonstrated by the increasing number of publications in which genitourinary malformations are no longer presented as individual case reports, but rather as reviews of large series (Hansmann et al. 1979; 1985; Weiss et al. 1981; Schmidt et al. 1981; Zerres 1981; Harrison 1983; Staudach et al. 1984; Hansmann 1984; Maurer et al. 1985). Descriptions of the normal renal morphology during gestation and corresponding biometric data have been published by Bernaschek and Kratochwil (1980), Grannum et al. (1980), Kratochwil (1982), Bowie et al. (1983), Jeanty and Romero (1984), and Hansmann et al. (1985).

9.1 Kidney

Estimates of the earliest gestational age at which the fetal kidneys can be visualized by ultrasound have been lowered in recent years, and we have been able to demonstrate them as early as 11–12 weeks given a favorable fetal position relative to the transducer (Figs. 9.1 a–c and 9.5).

After 15 weeks we were able to demonstrate fetal kidneys in 95% of cases with the use of 5-MHz transducers. Sonography of the fetal genitourinary tract is a classic example of the importance of using standardized planes of section.

9.1.1 Frontal Sections

Frontal views are generally best for evaluating the fetal genitourinary tract. The kidneys are located by first obtaining a frontal scan of the spine and then slowly moving the scan plane anteriorly until the spinal echoes disappear and the abdominal aorta comes into view (Fig. 9.2 a). At this level the kidneys appear relatively echo-dense in relation to their surrounding, while in all other planes the kidneys appear relatively sonolucent. Their apparent echogenicity in the frontal plane relates to the fact that the anatomic structures cranial to the kidneys are obscured by acoustic shadows from the ribs, giving this region a generally echo-poor ap-

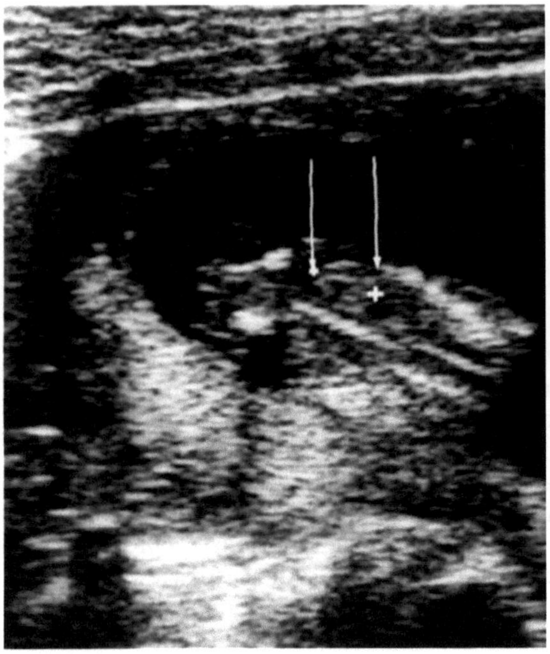

a

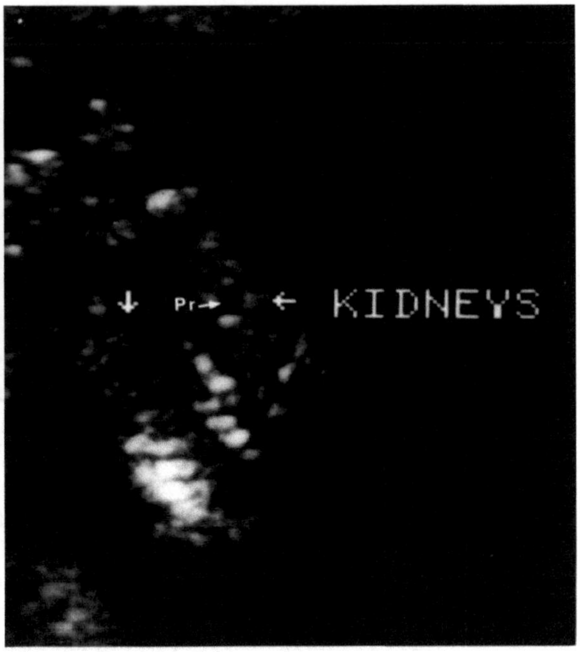

b

c

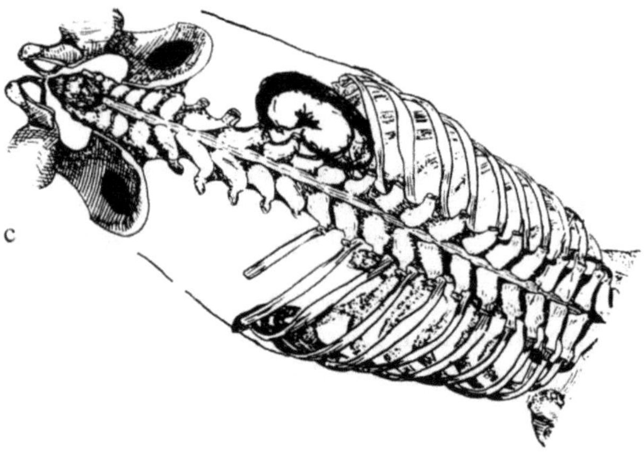

Fig. 9.1. a Early visualization of the kidneys (13 weeks) through ideal ultrasound windows. The beam in this frontal scan has free access to the kidneys through the space between the ilium and ribs

Fig. 9.1. b Visualization of the kidneys on a prevertebral frontal scan (fetal position as in Fig. 9.1 a) at 12 weeks. The renal parenchyma can be readily distinguished from the renal pelvis (*Pr*)

Fig. 9.1. c Schematic representation of the image plane

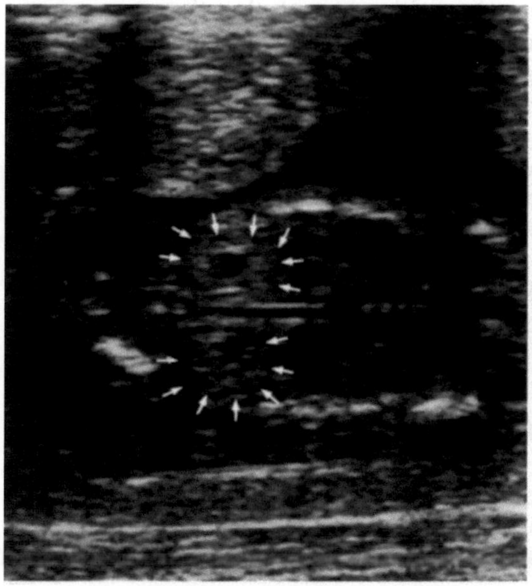

a

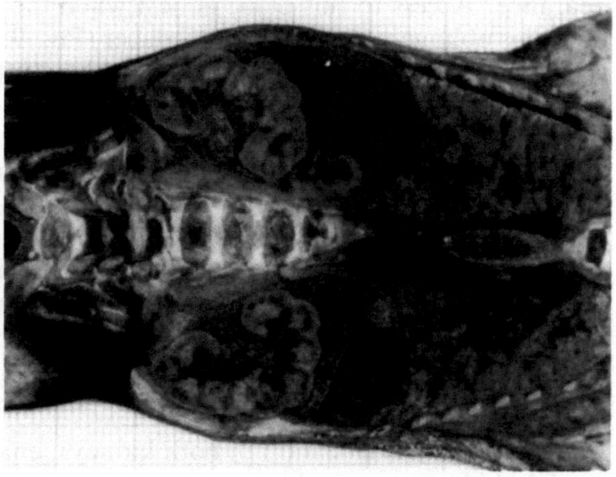

b

Fig. 9.2. a Visualization of both kidneys on a prevertebral frontal scan. The kidneys are positioned symmetrically to the right and left of the aorta, and the difference between the renal pelvis and parenchyma is apparent (15 weeks)

Fig. 9.2. b Frozen section from a 17-week fetus. This section also traverses both kidneys, but unlike the scan it additionally cuts the spinal column in the midline

Fig. 9.2. c Schematic drawing of the plane of section

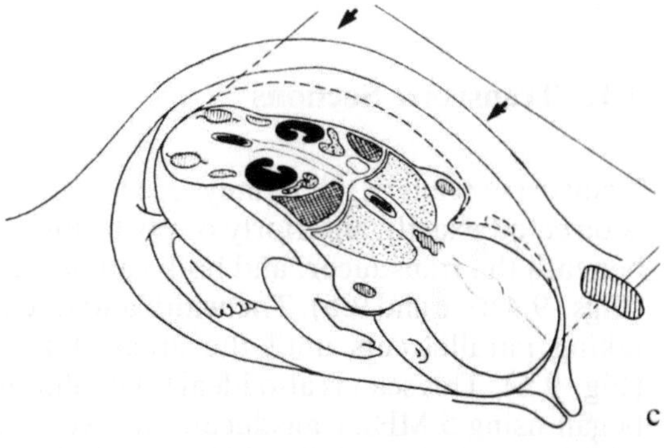

c

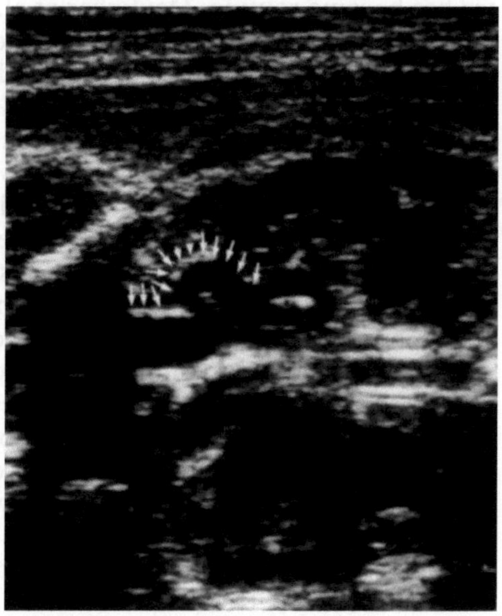

Fig. 9.3. Visualization of the ureter *(arrows)*, showing dilatation and tortuosity due to prevesical obstruction

pearance. Access to this plane is best when the fetus is positioned such that the body lies obliquely toward the transducer (Fig. 9.1 a–c). In that position the beam is able to penetrate the kidneys freely, without interference from the ribs or ilium, and in that position we were able to demonstrate the kidneys and renal pelves between 11 and 14 weeks.
A frontal scan is necessary for the concurrent evaluation of both kidneys and the urinary bladder, which can be visualized on one scan by angling the image plane slightly anteriorly in the region caudal to the kidneys. This is also the only plane in which we have been able to identify the ureters in cases of pathologic dilatation, which is understandable since this plane roughly follows the anatomic course of the ureters. Figure 9.3 shows ureteral dilatation and tortuosity associated with a prevesical obstruction.
We can summarize the importance of the frontal scan plane as follows:

1. Concurrent visualization of both kidneys, enabling comparison of the left and right kidney on a single scan.
2. Concurrent visualization of the kidneys and bladder.
3. Visualization of the ureters in cases of pathologic change.

9.1.2 Transverse Sections

Transverse scans of the kidneys are advantageous in fetal positions where the spine is directed exactly anteriorly or posteriorly. In these cases the kidneys lie parallel beneath the transducer, and both can be examined and compared on a single scan (Figs. 9.4 a–c and 9.5). The width and thickness of the kidney can be measured by taking parallel cuts until the greatest renal circular cross-section is visualized (Fig. 9.6). This scan is also ideal for evaluating renal pelvis size (Fig. 9.7). Until we began using 5-MHz transducers, we were rarely able to see the renal pelves. But with increasing use of these high-resolution scanners, visualization of the renal pelvis has become routine. Deutinger et al. (1984) examined the question of the degree of renal pelvic dilatation that should be considered pathologic and also calculated renal pelvic volumes. They found dilatations–defined as a maximum anteroposterior dimension of 5–8 mm–in 13% of their study population. They found no relationship between bladder filling and renal pelvis size. Maurer et al. (1985) described eight mild cases of renal pelvic distention in their series. In several of these cases, vesicoureteral reflux and mild prevesical stenosis was diagnosed postnatally. Hoddick et al. (1985) found fetal renal pyelectases exceeding 3 mm in 18%, noting that one-fourth of the ectasias were dependent on the state of maternal hydration. It may be concluded from these observations that dilatations of the renal pelvis exceeding 5 mm justify urgent and detailed ultrasound examination, and that postpartum examination by a pediatric urologist would be advisable. Our study population included a total of 11 cases of obstructive uropathy, and in every case the finding of renal pelvic distention prompted further examination that was able to establish the definitive diagnoses antenatally. The sonogram in Fig. 9.7 shows a case of subpelvic stenosis at 19 weeks, which was identified as such prenatally by repeat examinations. The psoas muscles may appear at this level as hypoechoic structures on the right and left side of the spine and should not be mistaken for dilated renal pelves (Fig. 9.8).

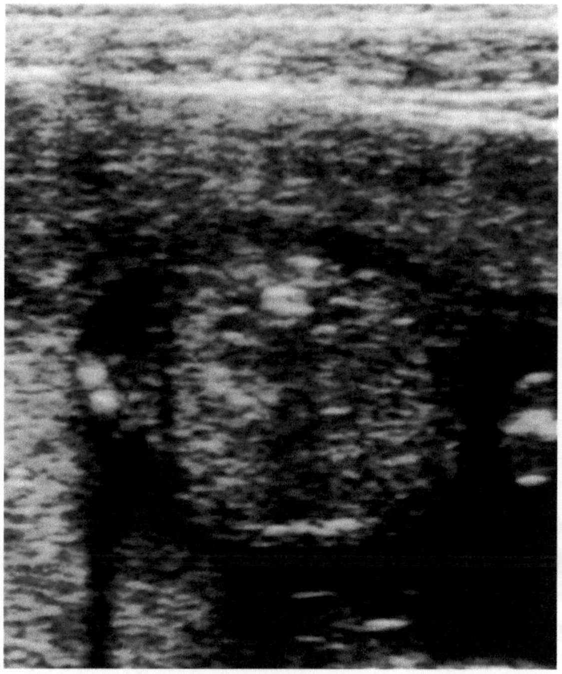

Fig. 9.4. a Visualization of the kidneys on a transverse scan. The fetus is prone, and the spine is flanked symmetrically by the kidneys. The renal pelves are visible in the central portion of the kidneys (15 weeks)

Fig. 9.4. b Frozen section in an analogous plane from a 21-week fetus. Fetal lobulation of the kidneys is apparent

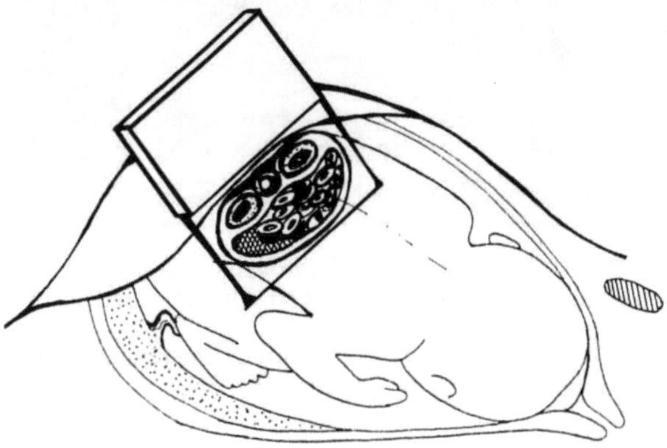

Fig. 9.4. c Schematic drawing of the plane of section

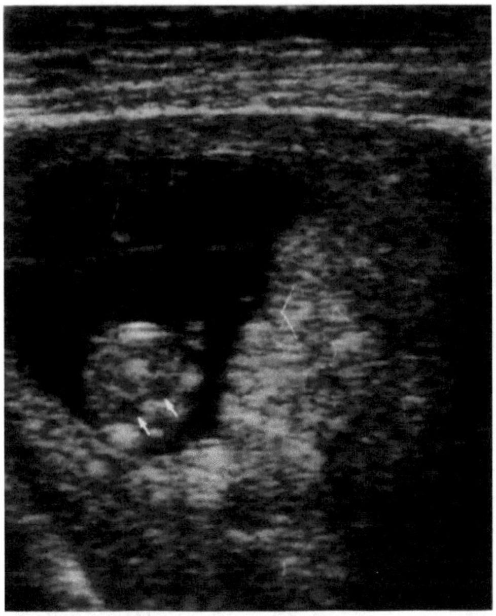

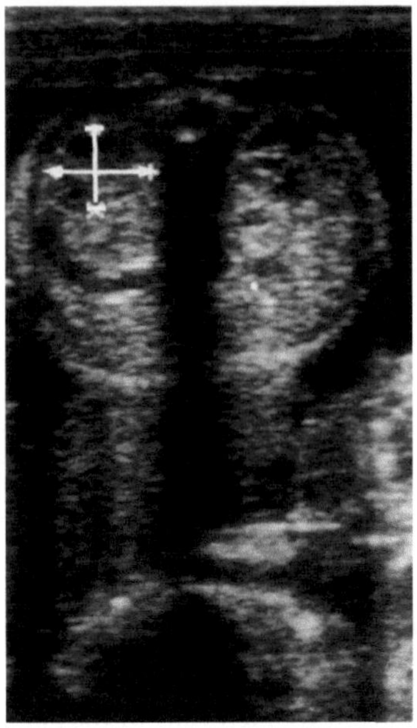

Fig. 9.5. (left) Early visualization of the kidneys on a transverse scan at 11 weeks. The kidneys *(arrows)* appear as circular structures adjacent to the spine, which is at 5 o'clock

Fig. 9.6. (right) Measurement of kidney width and depth on a transverse scan

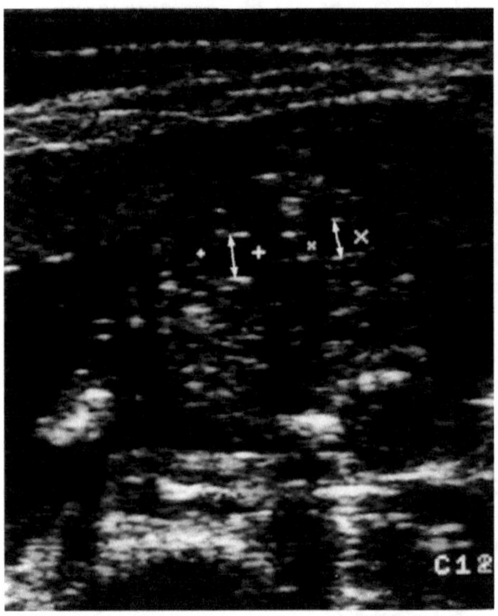

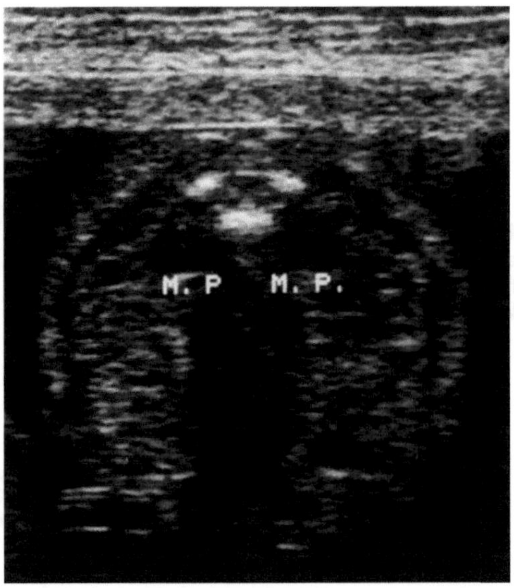

Fig. 9.7. (left) Transverse scan through the kidneys at 19 weeks. The renal pelves at the center of the kidneys are markedly distended. The AP diameter of the pelves is 9 mm

Fig. 9.8. (right) Transverse scan through the abdomen at 23 weeks. The hypoechoic paravertebral areas represent the psoas muscles (*M.P.*)

9.1.3 Sagittal Sections

To complete the screening of the fetal kidneys, the frontal and transverse scans are supplemented if possible by a paramedian AP sagittal scan (Fig. 9.10 a–c). This scan can display only one kidney at a time, so comparison of the kidneys requires that the transducer be moved across the midline to the contralateral side. This scan is particularly well suited for evaluating the topographic relationship of the kidney to the abdomen as a whole, and for measuring the kidney length (Figs. 9.9 and 9.10 a). The characteristic structural features of the fetal kidney can even be identified on a zoom image obtained with a high-resolution scanner (Fig. 9.13). It is possible to differentiate the pelvis, medulla, and cortex as well as the typical lobulation of the fetal kidney, analogous to an anatomic section (Figs. 9.11 and 9.13). Definitive studies on the development of the fetal kidneys were published by Oliver (1968), McCrory (1972), and Potter (1972).

9.1.4 Biometry

Accurate measurement of the transverse and longitudinal diameter of the fetal kidneys is warranted whenever there is suspicion of corresponding anomalies (Fig. 9.6). Nomograms were developed by Grannum et al. (1980) for determining the ratio of the kidney circumference to the abdominal circumference, and by Jeanty and Romero (1984) for determining the longitudinal, transverse, and anteroposterior diameter of the kidneys. It must be noted that the upper pole of the kidney cannot be clearly defined in all cases, because the adrenal gland is situated directly atop the kidney, and during fetal life this organ is quite large in relation to the kidney (Fig. 9.12). Especially when measurements are taken from a frontal scan (Fig. 9.14), the true upper pole of the kidney cannot always be clearly identified.

As gestation proceeds, the kidney shows an increasing degree of contrast with surrounding structures (Fig. 9.15), which Bowie et al. (1983) attribute to increased fat deposition in this area.

9.2 Adrenal Gland

Lewis et al. (1982) and Jeanty and Romero (1984) have reported ultrasound visualization of the fetal adrenal glands, and the latter authors have published tables of normal values. Figure 9.16 shows a transverse scan of the right adrenal gland, which is seen to consist of medulla and a sonolucent cortical rim. Our observations indicate that the right adrenal gland is more easily demonstrated than the left owing to its contrast with the liver.

We do not routinely image the adrenal gland because the diagnostic value of this structure is minimal at present. Perhaps new discoveries will be made one day that will justify ultrasound visualization of the adrenal gland as part of routine prenatal diagnosis.

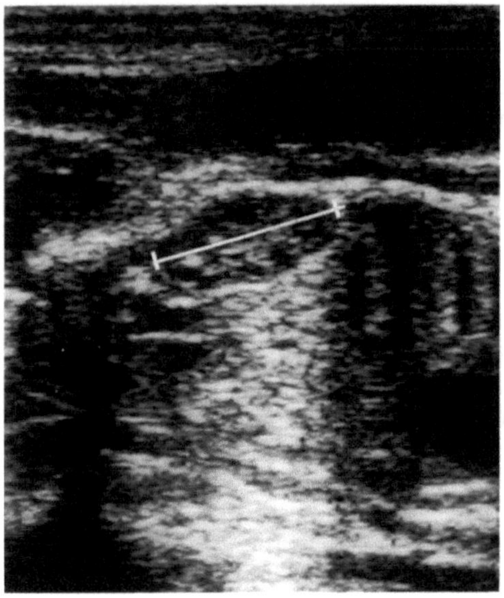

Fig. 9.9. Measurement of kidney length. This kidney is clearly delineated in the third trimester

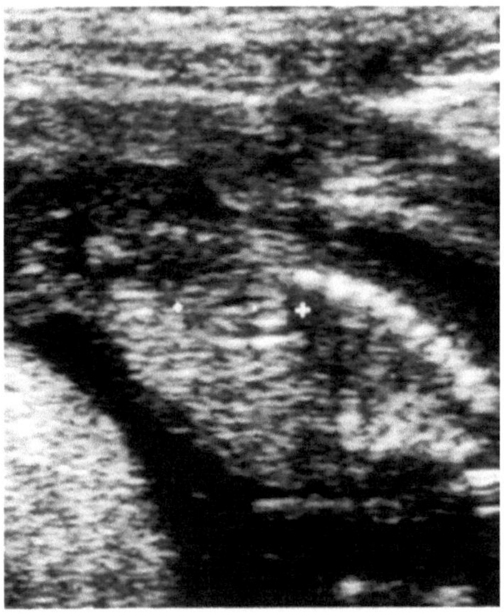

a

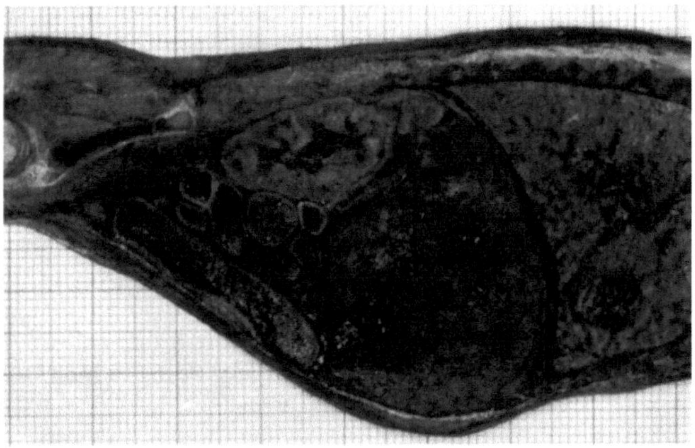

b

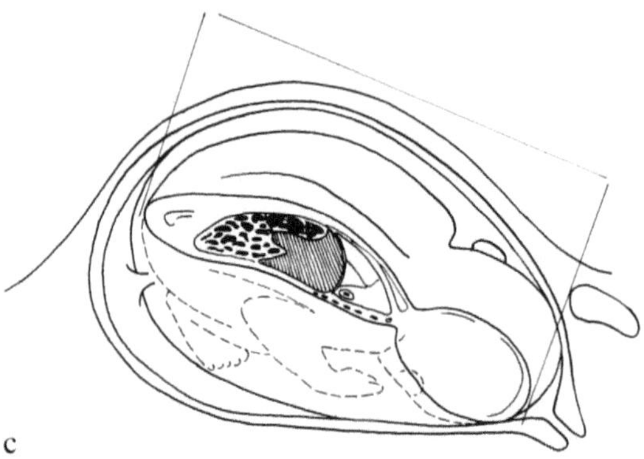

c

Fig. 9.10. a Visualization of the kidney on a paravertebral sagittal scan in a 16-week fetus (prone). The superior pole of the kidney is poorly defined for measurement of kidney length

Fig. 9.10. b Frozen section on an analogous plane from a 21-week fetus

Fig. 9.10. c Schematic drawing of the plane of section

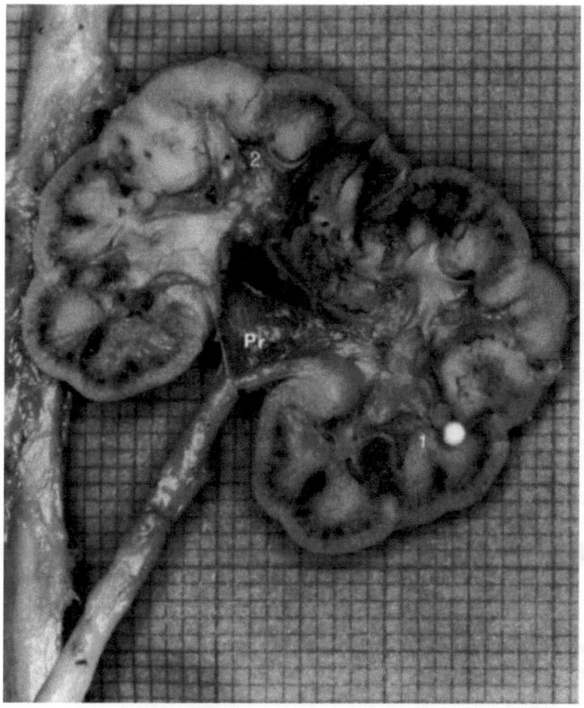

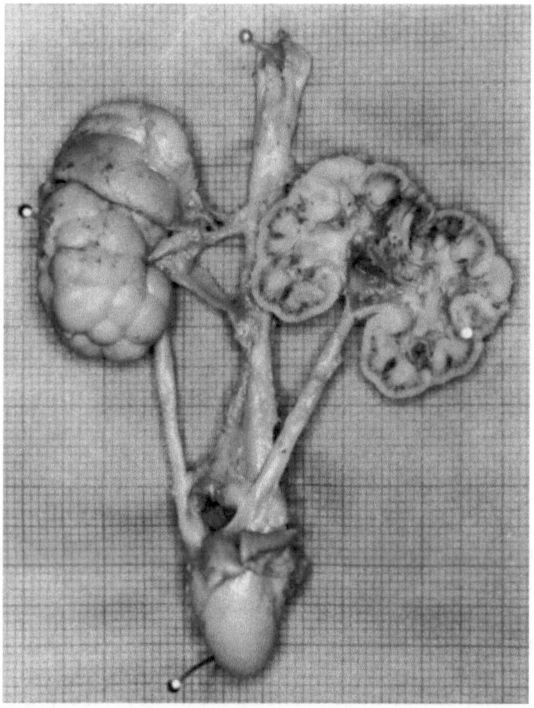

Fig. 9.11. (left) Frontal section through the center of a fetal kidney at 22 weeks, demonstrating the renal pelvis (*Pr*). Fetal lobulation is clearly visible. *1* Pyramids, *2* calyces

Fig. 9.12. (right) Genitourinary tract of a 22-week fetus. The ureters are dissected free, and the bladder is filled (urethra ligated). The aorta has been left attached to the specimen. The left kidney is sectioned to show its internal structure, and the large adrenal gland has been left in situ on the right kidney

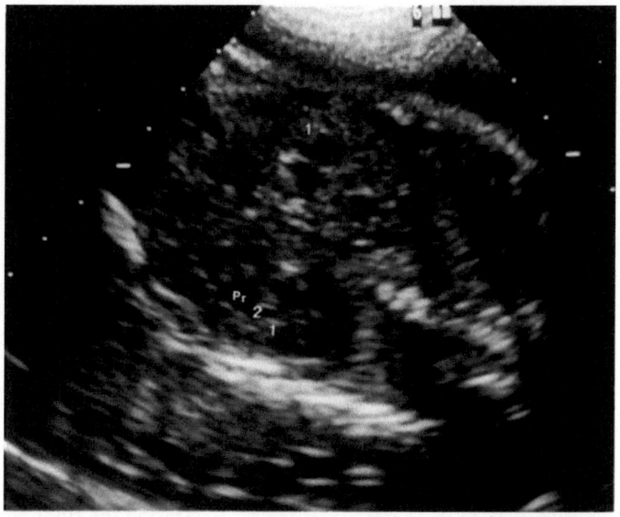

Fig. 9.13. Prevertebral frontal scan through the kidneys of a 27-week fetus. The renal pelvis (*Pr*) contrasts sharply with the renal parenchyma; the pyramids (*1*) and calyces (*2*) are demonstrated

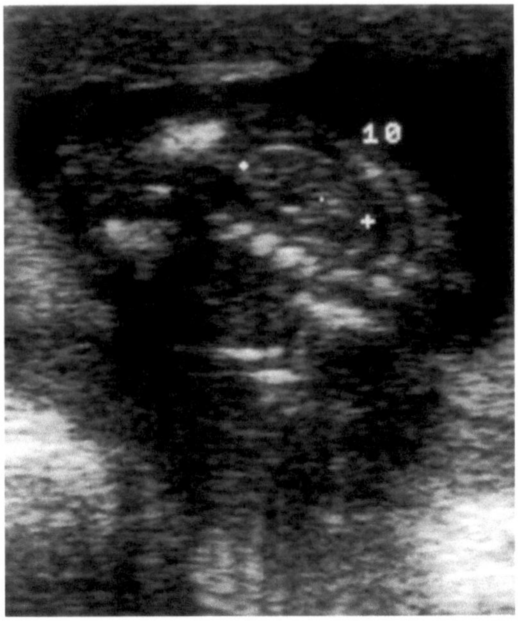

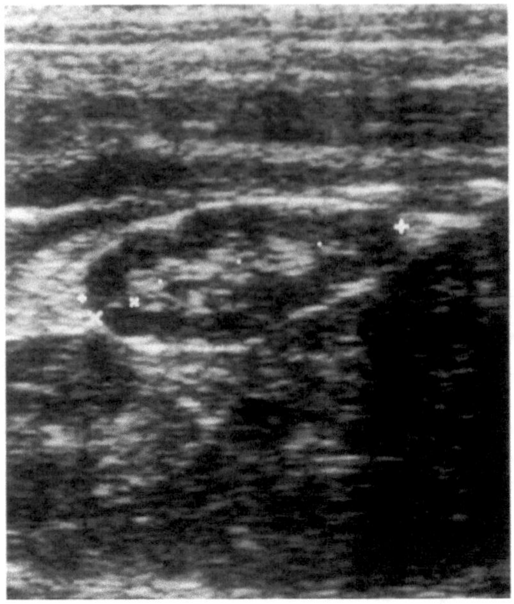

Fig. 9.14. (left) Visualization of the fetal kidney on a prevertebral frontal scan at 17 weeks. The superior part of the kidney is well demarcated from the adrenal gland for measurement of renal length. The tissue surrounding the kidney (fat capsule) is hypoechoic

Fig. 9.15. (right) Paravertebral sagittal scan through the kidney of a 39-week fetus. The kidney is not well demarcated from the adrenal gland for renal length measurement. The fetal lobulation is defined with great clarity by the contrasting, echogenic capsule

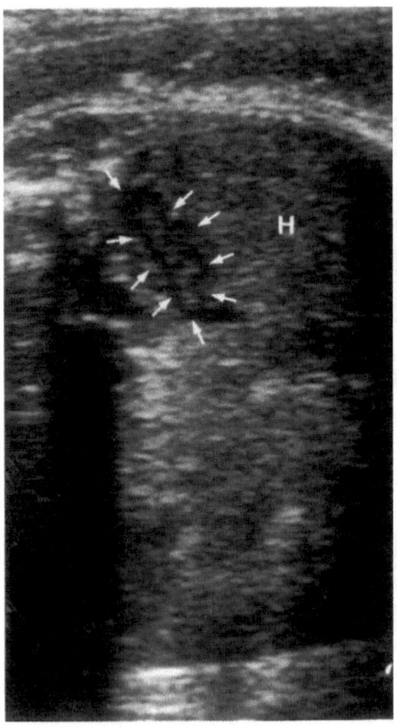

Fig. 9.16. Transverse scan in a 39-week fetus demonstrating the right adrenal gland. The adrenal gland *(arrows)* appears hypoechoic in relation to the liver, and the medulla is barely distinguishable from the cortex

9.3 Bladder

A meaningful evaluation of the genitourinary tract is possible only by combining assessment of the amniotic fluid volume (see Chap. 3) with a concurrent evaluation of the fetal urinary bladder. Normally the bladder appears as a round to oval, cystic structure in the midportion of the lower abdomen (Fig. 9.17). However, this structure can be confidently identified as the bladder only if it demonstrates rhythmic filling and emptying. In our screening population, a fatal misdiagnosis resulted from the misinterpretation of a cystic kidney as the urinary bladder (see Sect. 2.2). If doubt exists as to the identity of a fluid-filled cavity in the lower abdomen, it is best to perform serial examinations.

Our present knowledge of the physiology of the fetal bladder stems largely from the work of Campbell et al. (1973), Wladimiroff (1978), Wladimiroff and Campbell (1974), Wladimiroff et al. (1976), Visser et al. (1981), and Kurjak et al. (1981). The bladder volume (Campbell et al. 1973) is determined by measuring the maximum bladder dimensions in length, width, and depth and applying the formula $V = 0.52$ abc to calculate the volume. Using this technique, it has been shown that the bladder empties cyclically with a period of 50–155 min, with the emptying phase occurring very rapidly. Brusis et al. (1975) reviewed the dynamics of amniotic fluid volume changes associated with fetal urine production. We personally observed that fetal bladder emptying can be stimulated by percussing the uterine wall. Details on diagnostic procedures that can be implemented in patients with oligohydramnios and a suspected genitourinary anomaly (amniotic fluid replacement, pharmacologic diuresis) have been presented by Hansmann et al. (1985) and are outside our present scope. Hansmann et al. (1985) described a typical "jet sign" in the amniotic fluid that accompanied fetal bladder emptying, but the diagnostic value of this phenomenon does not justify the time required to document it. In the course of our own examinations, we found that the fetal urethra could be delineated during urination in males and females (Figs. 9.18 and 9.19), although these are fortuitous findings that have no real bearing on routine diagnosis.

9.4 Genitalia

Besides the technical questions of earliest detection time and criteria for recognizing the fetal genitals by ultrasound, the subject of antenatal sex diagnosis has given rise to a number of ethical, psychological, and legal questions. A detailed review of the current status of the literature on fetal sex identification was published by Eleyalde et al. (1985). These authors reviewed a total of 3891 cases published by numerous authors, noting the accuracy of sex identification in relation to gestational age and gender. Analysis of the survey showed that fetal gender could be correctly determined in 65.6% of cases.

The two studies with the largest populations (Birnholz 1983; Eleyalde et al. 1985) indicate a 40% and 60% rate of visualization of the fetal genitalia, respectively, before 24 weeks, with an error rate of 3%. The earliest identification was at 12 weeks. Except for the study of Stephens and Sherman (1983), who reported a 100% accuracy rate in 100 cases, no series have been free of error. This fact should be

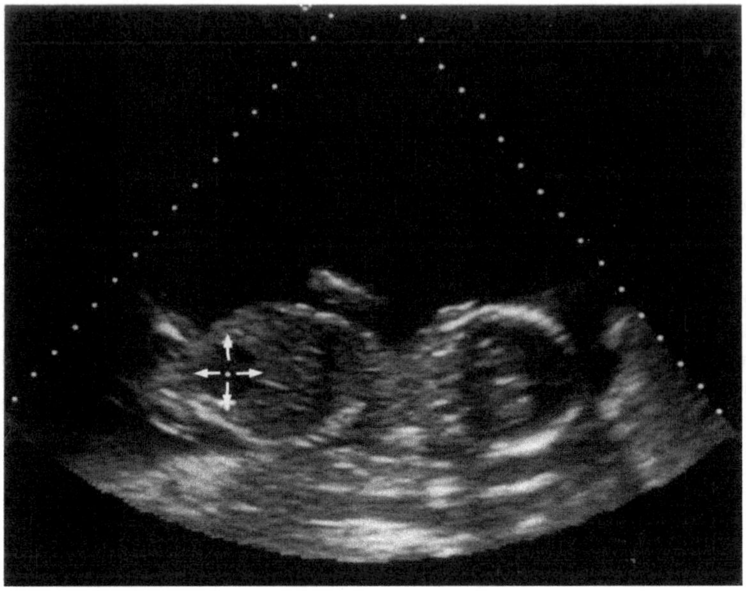

Fig. 9.17. Visualization of the bladder in a 19-week fetus; pronounced hydramnios. The *arrows* mark two of the three parameters used to calculate bladder volume

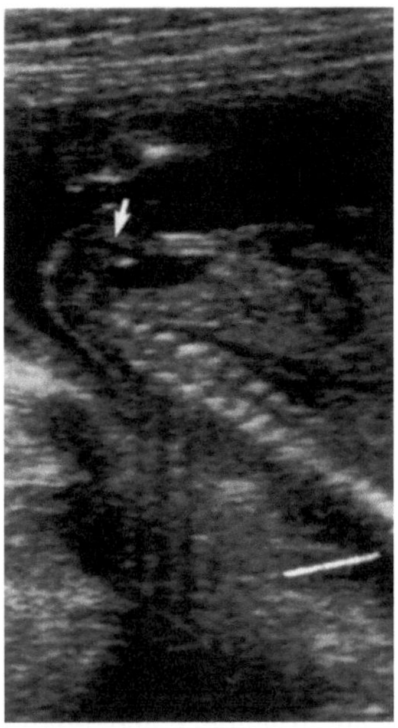

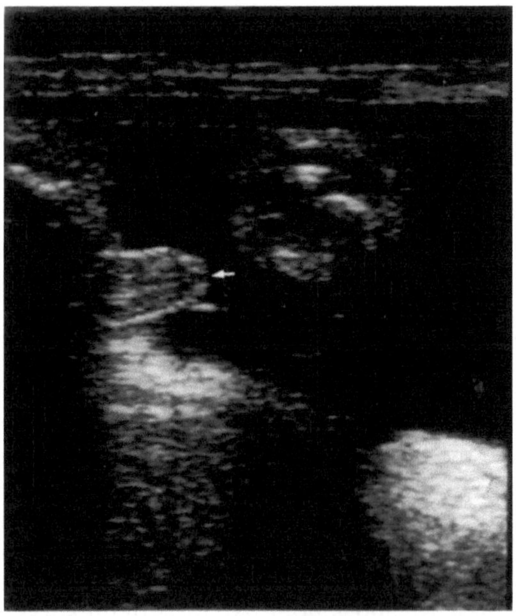

Fig. 9.18. (left) Midsagittal scan in a female fetus at 19 weeks. The urethra *(arrow)* is visible during urination

Fig. 9.19. (right) Scan through the penis of a 33-week fetus. The urethra *(arrow)* is visible as a dark, central band during urination

emphasized, because every antenatal sex identification by ultrasound carries a risk of misinterpretation in the absence of other tests. Thus, in all cases where sex identification may have implications from a genetic standpoint, karyotyping should be performed before definitive conclusions are drawn. This uncertainty factor supports existing legislation in Austria which frees the physician from an obligation to inform the prospective parents of the fetal gender–unlike the situation in the United States where legal decisions appear to mandate such a disclosure (Kass and Shaw 1976; Kenna 1973). The ethical and psychological dilemma posed by sex disclosure is illustrated in a case described by Stenchever (1972). The prospective mother desired to have amniocentesis, ostensibly for the purpose of excluding Down's syndrome. When the test showed a genetically normal female, the woman chose to have the pregnancy terminated.

However, there are several medical arguments which support ultrasound sex diagnosis:

1. The possibility of detecting testicular feminization (Stephens 1984)
2. The confirmation of correct amniocentesis technique in cases of multiple gestation (Eleyalde and DeEleyalde 1984)
3. Support for genetic laboratories in the confirmation of maternal cell contamination if female karyotype was found and a male fetus diagnosed by ultrasound.
4. The direct demonstration of genital malformations (Hansmann et al. 1985; Cooper et al. 1985)

These considerations alone prompted us to address the possibility of fetal sex identification from an anatomic standpoint in this volume. The earliest time at which we could determine fetal gender in our series was at 12 weeks (Fig. 9.20). It should be emphasized, however, that we made this determination using a high-resolution scanner of the latest generation. Based on the factors cited earlier, we have not undertaken a prospective study of antenatal sex determination, and so we are unable to present data on the quality of our performance in this regard. Previous experience indicates that it is easier, to establish a male fetal gender than a female. This is consistent with the results of most authors. Sonographic demonstration of the male genitalia is most easily accomplished by an AP midsagittal scan (Figs. 9.20 and 9.21) or in cases where the legs are slightly abducted, by a tangential scan of the caudal fetal pole. In a few cases we have documented fetal penile erection (Fig. 9.22), which is consistent with the observations of Eleyalde et al. (1985).

The female genitalia are most easily identified on tangential scans with the legs drawn up. Differentiation of the labia majora and minora is possible in some instances (Fig. 9.23).

Evidence for the possibility of detecting genital anomalies by ultrasound as one aspect of a syndrome or chromosomal disorder (Hansmann et al. 1985) justifies the acquisition of diagnostic experience in anatomically normal fetuses to develop the necessary basis for qualitatively accurate assessment in high-risk cases. The prerogative of the physician to disclose or withhold information on the fetal gender should, in our view, be respected.

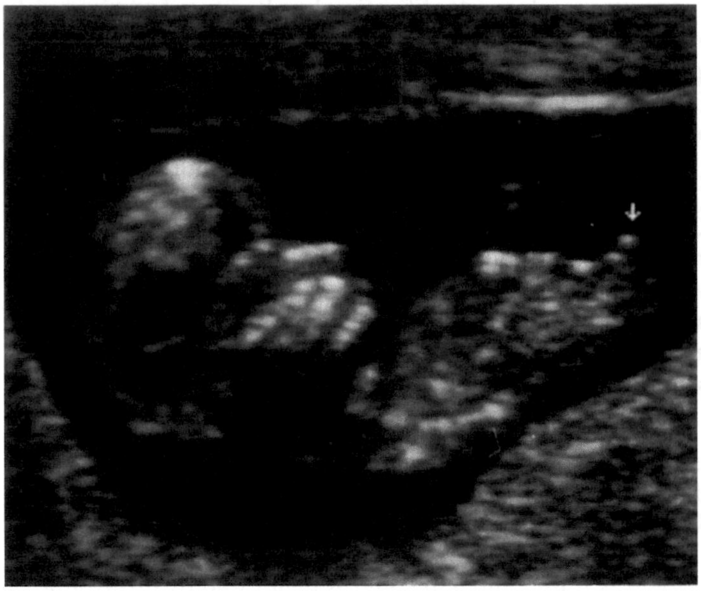

Fig. 9.20. Sagittal scan of a supine fetus at 12 weeks. The penis is visible at the caudal end of the trunk *(arrow)*

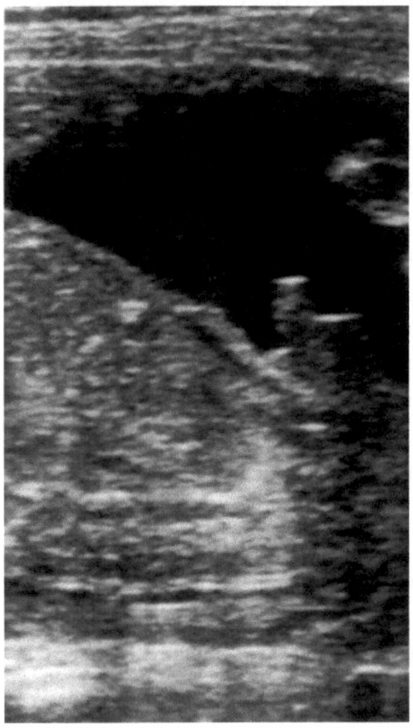

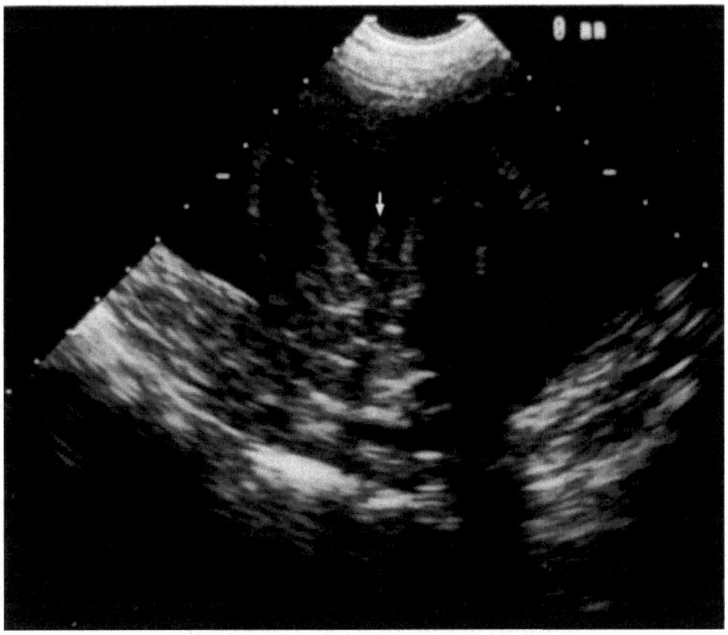

Fig. 9.21. (left) Midsagittal scan of a male fetus at 21 weeks. The image plane traverses the penis and scrotum anteriorly

Fig. 9.22. (right) AP transverse scan through the fetal lower abdomen and proximal lower limbs. The erect penis *(arrow)* is visible between the abducted legs

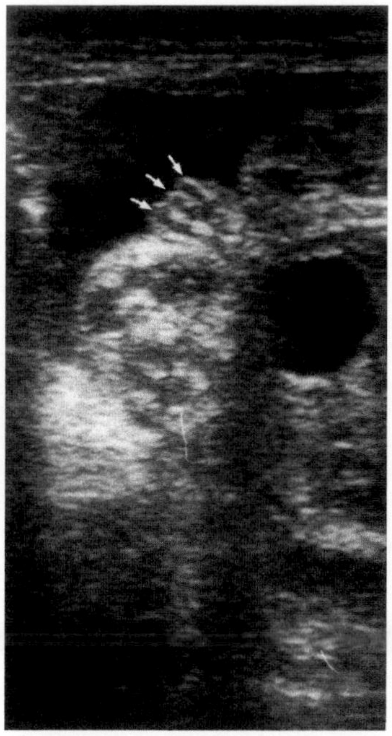

Fig. 9.23. Visualization of the female genitalia on a tangential scan of the caudal fetal pole. The labia majora and labia minora *(arrows)* can be identified. The bladder is full

10 Skeleton

Because of their special characteristics and anatomic relationship, the osseous structures of the fetal head and spine have been described separately. The present chapter deals with the sonographic features of the sternum, ribs, clavicle, scapula, upper extremity, pelvis, and lower extremity. As these bones acquire ossification centers, they become increasingly accessible to sonographic examination. Purely cartilaginous structures also can be delineated by ultrasound (e.g., the condyles of the long bones, Figs. 10.16 and 10.31), but in the earlier stages only the ossified portions of the fetal skeleton are directly visible. Table 10.1 shows a list of the individual bones along with the earliest times at which they could be detected in our studies. It should be noted that the times of appearance of these bones become earlier with each new generation of scanner. As resolution improves, it becomes possible to identify areas of very low sonodensity, as long as their acoustic impedance contrasts with that of surrounding tissues. Thus, data on the times of appearance of ossification centers on sonograms are in no way comparable to

Table 10.1. Earliest sonographic detection of ossification centers in the fetal skeleton

Clavicle	8–9 weeks
Humerus Radius Ulna Femur Tibia Fibula	9 weeks
Scapula Ribs Ileum	10–11 weeks
Fingers Toes	11–12 weeks
Ischium	16–17 weeks
Pubis Calcaneus Talus	19–23 weeks
Sternum	21–27 weeks
Epiphysis of femur	29 weeks
Epiphysis of tibia	35 weeks

radiologic data. Fetal ossification centers provide conspicuous landmarks for sonographic examination. Knowledge of the times of appearance of the different ossification centers and of their topographic relations and dimensions provide an important basis for orientation. Skeletal structures are also gaining increasing importance as sources of biometric data.

10.1 Sternum

The sternum is formed from bilateral sternal plates which chondrify and begin to fuse with the ribs at about 10 weeks' gestation. This bone shows extreme individual variation in its development and especially in its ossification centers, which are highly variable in their location and time of appearance. It is these ossification centers that are imaged by ultrasound. Even in the second trimester, the sternal ossification centers appear only as spotty echo densities in the sternal cartilage (Fig. 10.1). Appreciable coalescence of these areas is not evident until birth. They can be imaged sonographically in all three scanning planes and appear as individual sonodense areas (Figs. 10.2 and 10.3). A frontal, tangential scan is required to demonstrate their "string of beads" arrangement along the course of the sternum (Fig. 10.4). The areas between the ossification centers can be used as windows for scanning intrathoracic structures; inability to visualize the entire sternum should not be considered abnormal.

10.2 Ribs

Ribs generally can be visualized after 9–10 weeks and provide for an increasingly clear delineation of the fetal thorax. By 12 weeks as many as 11 ribs can be demonstrated under favorable conditions (Fig. 10.5). Especially on symmetrical frontal scans, the ribs appear as echogenic band like structures projecting in a fanlike pattern from the lateral ossification centers of the vertebral arches (Fig. 10.6). The sonographic importance of the ribs is in evaluating the overall shape and symmetry of the thorax. Imaging the 12th rib on a frontal scan can be helpful in the localization of neural tube defects, as it provides a landmark for counting the ossification centers in the lumbar spine (Fig. 10.7). Rib imaging also can disclose costal dysplasias, which occur in a number of osseous syndromes (Smith 1982).

10.3 Clavicle

The clavicle is the first bone in which ossification centers appear. As yet, however, this circumstance has no significance from the standpoint of ultrasound diagnosis. At the end of the first trimester, the clavicle is always visible as a high-level echo at the junction of the neck and thorax (Fig. 10.8). It provides a useful landmark for locating anatomic structures in this region, especially on AP sagittal scans (Fig. 10.9). Yarkoni et al. (1985) performed serial measurements of the clavicle

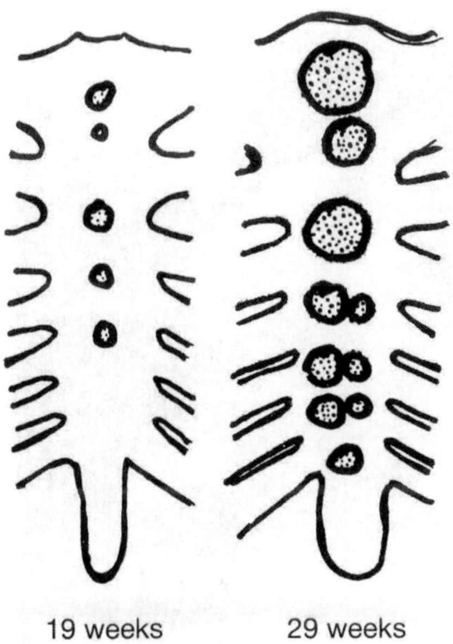

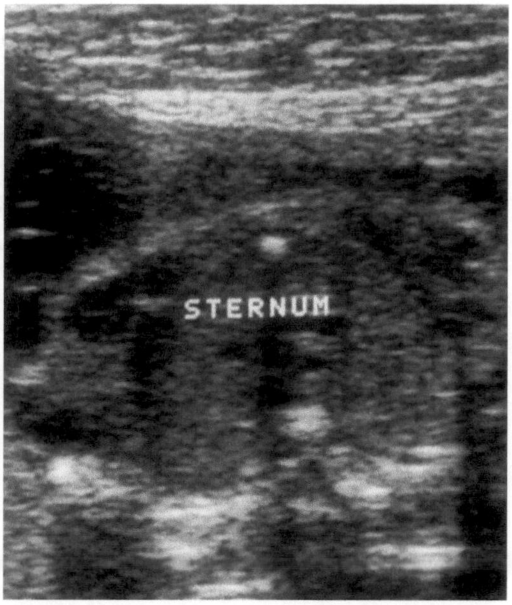

Fig. 10.1. (left) Schematic drawing of the ossification centers of the sternum at 19 and 29 weeks

Fig. 10.2. (right) Transverse scan through the thorax of a 27-week fetus. The scan is placed through one of the sternal ossification centers

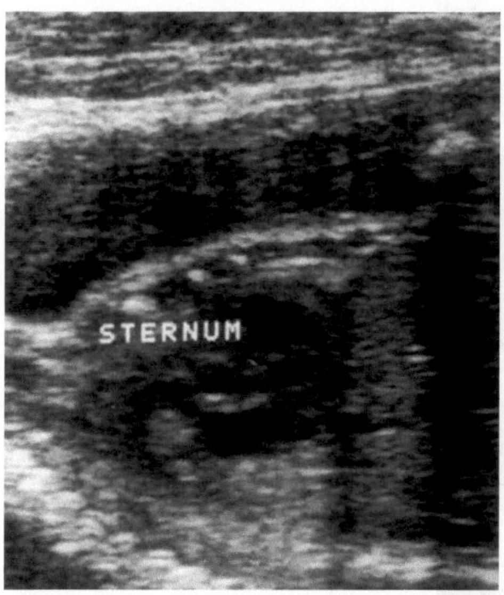

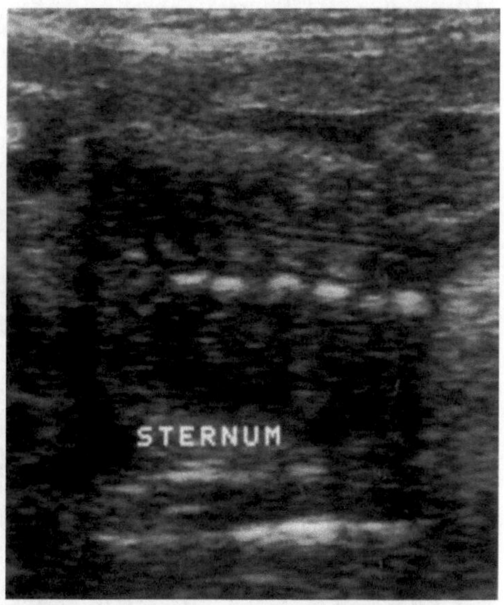

Fig. 10.3. (left) Midsagittal scan through a fetal thorax at 27 weeks. Several ossification centers are visible in the sternum

Fig. 10.4. (right) Tangential, anterior frontal scan through the chest of a 27-week fetus. The sternal ossification centers display a "string-of-beads" configuration

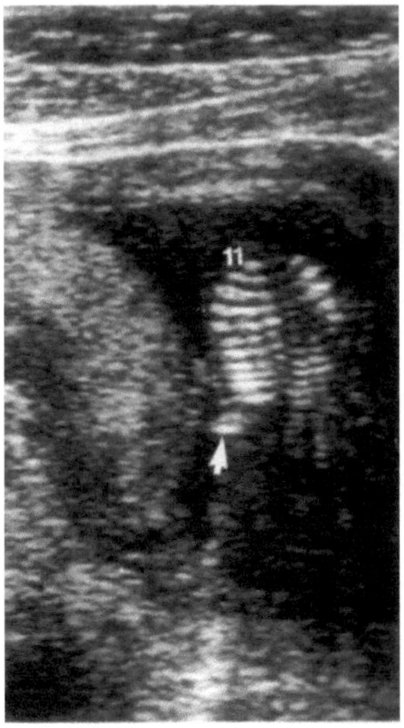

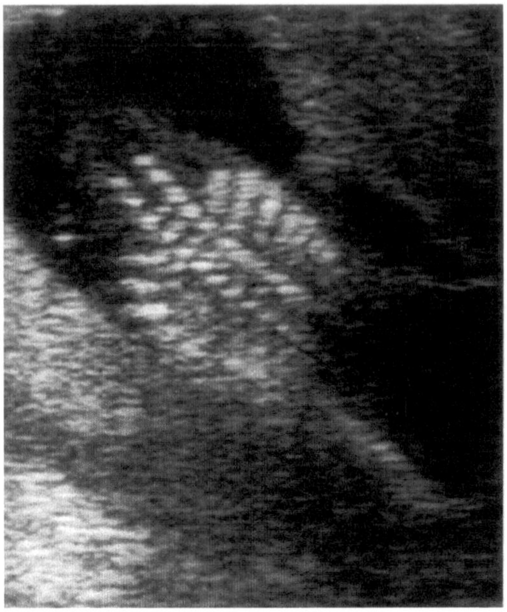

Fig. 10.5. (left) Ultrasound appearance of the fetal ribs at 12 weeks. The scan excludes the head, which is directed caudally. A total of 11 ribs are visualized. The *arrow* marks the scapula

Fig. 10.6. (right) Symmetrical frontal scan through the thorax at 14 weeks. The ribs project symmetrically from the lateral ossification centers of the thoracic vertebral bodies

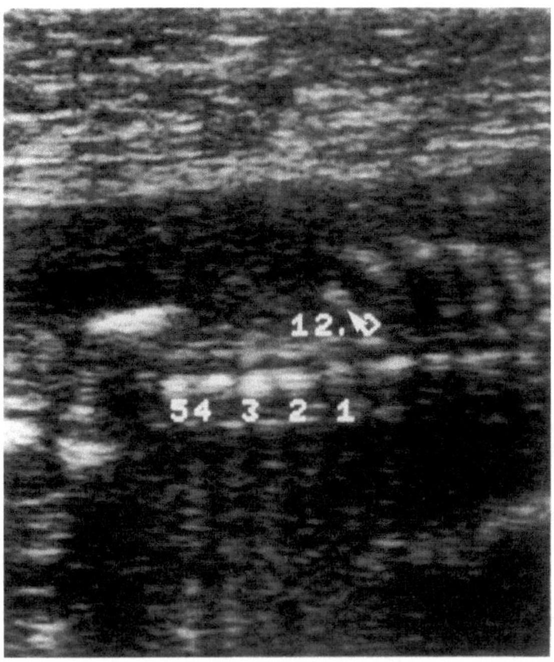

Fig. 10.7. Frontal scan. The 12th rib provides a reference point for counting the ossification centers of the lumbar vertebrae

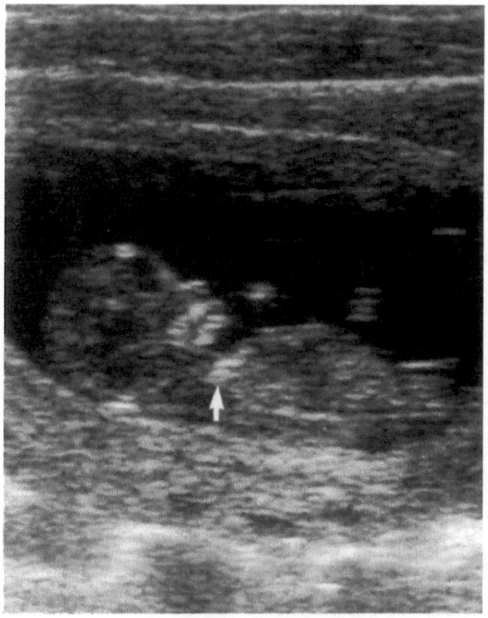

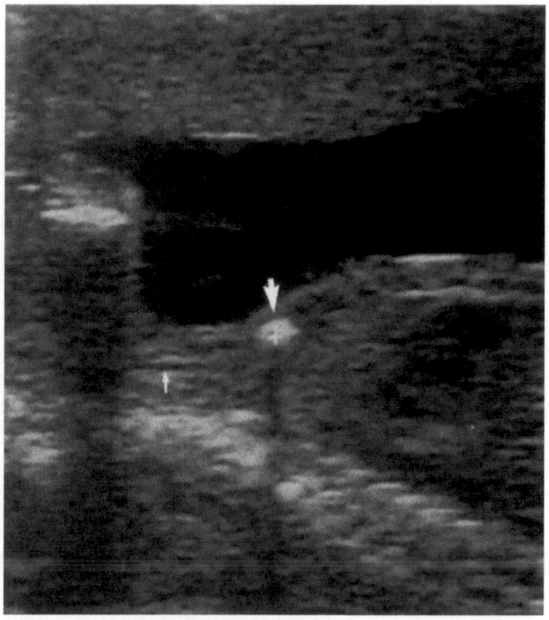

Fig. 10.8. (left) AP parasagittal scan of a 12-week fetus. The clavicle, cut transversely on this scan, appears as a bright echo *(arrow)* at the junction of the neck and thorax

Fig. 10.9. (right) Ap scan through the fetal neck and thorax. The scan displays the carotid artery *(small arrow)* region and a cross section through the clavicle *(large arrow)*

and noted a linear correlation with gestational age. Besides this additional parameter for assessing fetal age, measurement of the clavicle is also important for the specific exclusion of some syndromes (cleidocranial dysostoses, Goltz syndrome, Holt-Oram syndrome, Melnick-Needles syndrome; survey in Smith 1982). The clavicle can be visualized and measured on transverse scans in the AP or PA position (Figs. 10.10 and 10.11). It is necessary to image the full length of the clavicle, and this is most easily accomplished on a symmetrical scan that displays both clavicles on one image (Fig. 10.11). The bone ends are easily identified, because the neighboring parts of the scapula and sternum contain no directly adjacent ossification centers that might obscure the clavicle borders.

10.4 Scapula

The scapula can be demonstrated in isolation after 12 weeks on tangential scans of the posterior body surface (Fig. 10.12), where it may appear patchy or ringlike depending on the tangential position of the beam. From about 15–16 weeks the scapula displays its characteristic shape on sonograms (Fig. 10.13), although the visible area by no means corresponds to the total anatomic size of the bone. At present, the scapula has importance on sonograms only as a landmark and also as a means of confirming the symmetry of the image plane when both scapulae are demonstrated on a frontal scan.

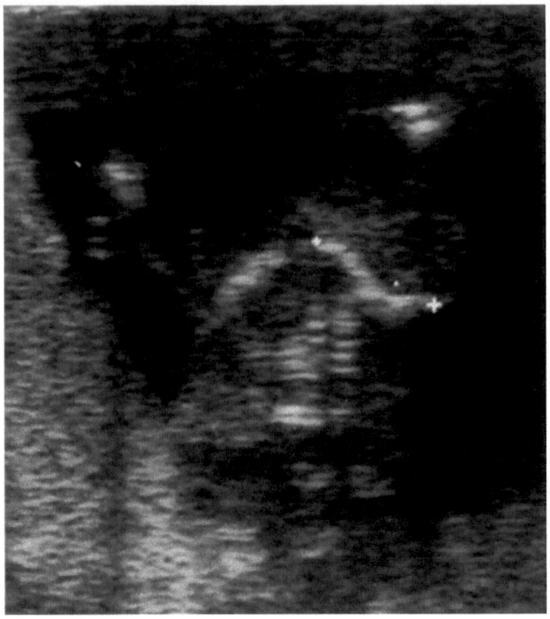

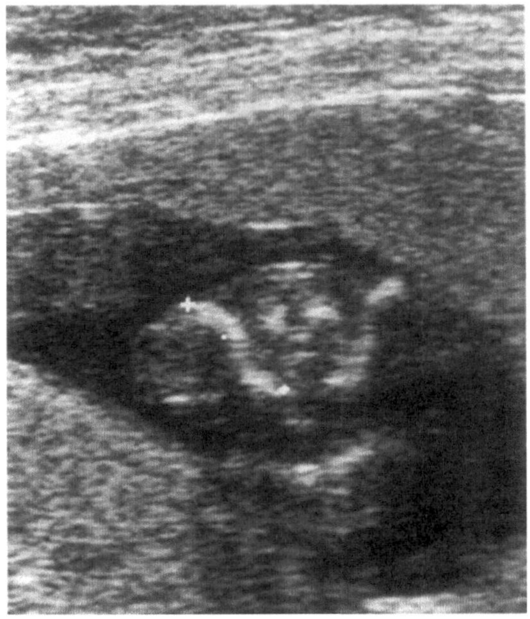

Fig. 10.10. (left) Measurement of clavicle length in a 14-week fetus on a transverse scan. The spine is at 6 o'clock

Fig. 10.11. (right) Measurement of clavicle length in a 16-week fetus. The fetus is lying prone relative to the scanner. The simultaneous visualization of both clavicles confirms the symmetry of the scan

10.5 Pelvis

Given the varying ossification times of the individual pelvic bones, the pelvis cannot be imaged in its entirety before 19 weeks. Pelvic ossification begins in the ilium, and the iliac centers can be observed by ultrasound at about 10–11 weeks. On frontal scans performed through the spine, the iliac wings appear bilaterally as bright, oval-shaped echoes (Fig. 10.22). We have found that the ossification centers of the ischium can be detected at about 15–16 weeks, appearing on frontal scans as rounded echogenic structures at the caudal end of the pelvis, separate from the ilium. Their size increases rapidly with gestational age (Fig. 10.23 a, b). In one case we were able to identify the pubic bone before 21 weeks. This was accomplished on a transverse scan through the caudal pelvis (Fig. 10.24 a, b). The pubic ossification centers appear as two separate echoes medial and caudal to the ischial centers on a PA transverse scan. These structures serve primarily as landmarks for locating the proximal portions of the lower extremities. They also produce acoustic shadows in scans of the lower abdomen, which must be taken into account when differentiating normal from abnormal findings.

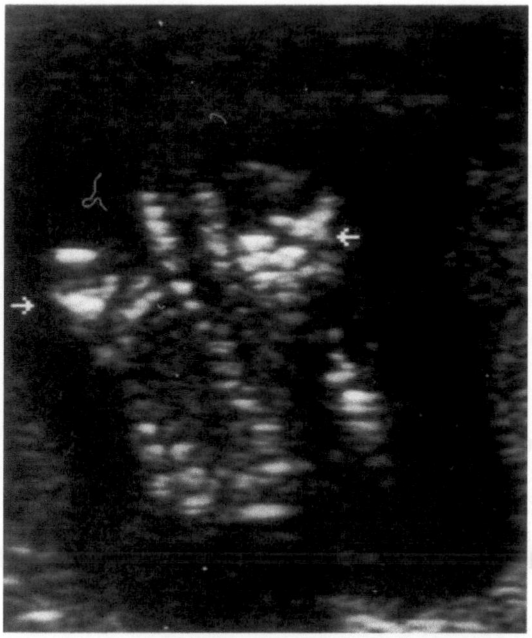

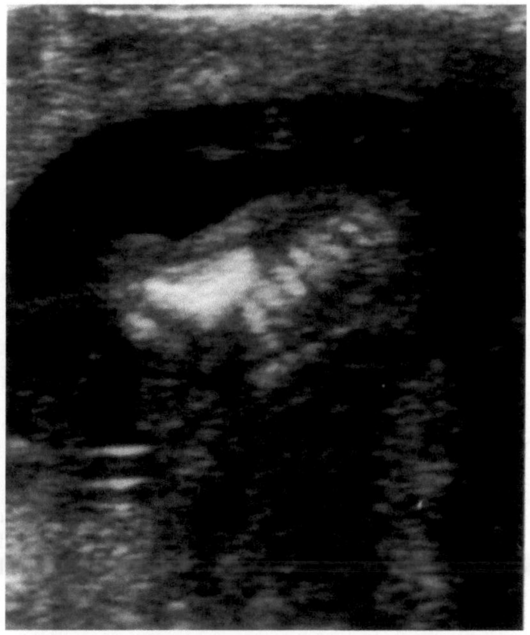

Fig. 10.12. (left) Frontal scan through the back of a 12-week fetus. The scan misses the head, which is directed upward to the left. The ossification centers of both scapulae are seen adjacent to the spine. The concurrent visualization of both scapulae confirms a symmetrical scan

Fig. 10.13. (right) Visualization of a scapula on a tangential frontal scan at 16 weeks. The head (missed by the scan) is directed downward to the left

10.6 Extremities

Examination of the fetal extremities currently is not a part of routine scanning, although the time required to visualize the extremities has been greatly reduced by the use of real-time equipment (Schlensker 1982). Questions concerning the presence and integrity of the fetal extremities are probably the most frequent asked by prospective mothers. In this respect the examiner can perform a meaningful psychoprophylactic service by locating the extremities and demonstrating them to the patient. Increasingly, we do this even in routine examinations. With experience, and by adhering to principles that greatly facilitate the examination procedure, the examiner can demonstrate the fetal extremities in far less time than is generally assumed.

10.6.1 Technique

We have had good results with the following procedure (Fig. 10.14 a, b): First the proximal part of the limb is located on a transverse scan. Using that area as a pivot point, the transducer is rotated until the shaft of the humerus of femur comes into view. In the next phase, the elbow or knee joint serves as the pivot point about

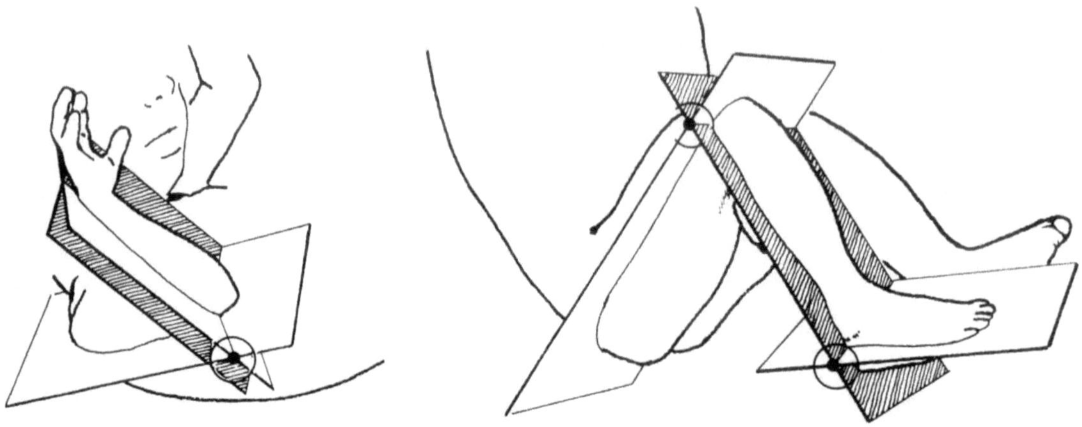

Fig. 10.14. a Schematic drawing of the technique used for detailed evaluation of the upper extremity

Fig. 10.14. b Schematic drawing of the technique used to locate the parts of the lower extremity. After the femur is identified, the transducer is rotated about the knee joint to image the long axis of the lower leg. Similarly, the transducer is rotated about the ankle joint to demonstrate the sole of the foot

which the transducer is rotated to visualize the distal parts of the extremity. To find individual bones (radius and ulna, fibula and tibia), the examiner maintains the basic direction of the scan while rotating the transducer about the long axis of the extremity. The same technique can be applied in the ankle area to demonstrate the sole of the foot and the toes. The technique is somewhat more difficult in the hand, since a common image plane rarely exists beyond the wrist due to the mobility of the digits.

As early as 1972, Holländer noted the advantages of real-time scanning in examinations of the fetal extremities. Hoffbauer et al. (1978) performed a comprehensive study on measurements of the fetal extremities, while Mahony and Filly (1984) reviewed the diagnostic capabilities of high-resolution scanners in this area. The importance of fetal limb imaging in the diagnosis of malformations has been stressed in numerous publications (Richardsen et al. 1977; Mahoney and Hobbins 1977; Luthy et al. 1979; Lang et al. 1979; Hobbins and Mahoney et al. 1980; Smith et al. 1982; Filly et al. 1981; Staudach et al. 1982; Hobbins et al. 1982; Winter et al. 1985; Jeanty et al. 1985b).

10.6.2 Biometry

Until 1980, measurement of fetal head and trunk parameters constituted the focus of ultrasound biometry. Since that time, increasing importance has been ascribed to measurements of the long bones of the extremities (Queenan et al. 1980; Hoffbauer 1981; Hohler and Quetel 1981; O'Brien et al. 1981; Schlensker 1981; Terinde et al. 1981; Yeh et al. 1981; O'Brien and Queenan 1981, 1982; Hohler and Quetel 1982; Hadlock et al. 1982, 1983a,b, 1984; Jeanty et al. 1984; Hansmann et al. 1985). Of the many parameters that have been described, measurement of the femur length is considered the most useful, and we routinely determine this parameter even in basic examinations. Lately there have been reports of discrepancies in femur length measurements performed with linear and sector scanners (Hills et al.

1982; Leo et al. 1983a,b; Pretorius et al. 1983; Winsberg 1983). Jeanty et al. (1985a) in experimental studies found deviations of up to 14% from the true femur length using sector and linear scanners made by various manufacturers, although the average deviation, at 4%, was within acceptable limits. Nevertheless, this point should be given more serious consideration in the future and may account for the discrepancies in the data reported by different authors.

10.6.3 Upper Extremity

The upper extremity is generally more difficult to evaluate by ultrasound than the lower extremity, because the fetus tends to move it more frequently and with greater amplitude. Simultaneous visualization of the humerus, radius, ulna, and fingers is unusual (Fig. 10.15) and possible only when the fingers are extended and the entire extremity is imaged in one plane. As the figure indicates, only the ossified portions of the extremity bone are visible. The elbow joint itself and the carpal region appear "vacant" except for some low-level soft-tissue echoes. Thus we can easily identify the distal, middle, and proximal phalanges and the individual bones of the metacarpus.
Visualization and measurement of the diaphyses generally should be done on the extremity that is closer to the transducer, and the limb orientation should be parallel to the transducer. The measurements at 18 weeks in Figs. 10.16 and 10.17 show how clearly the points of measurement are defined when the transducer is correctly

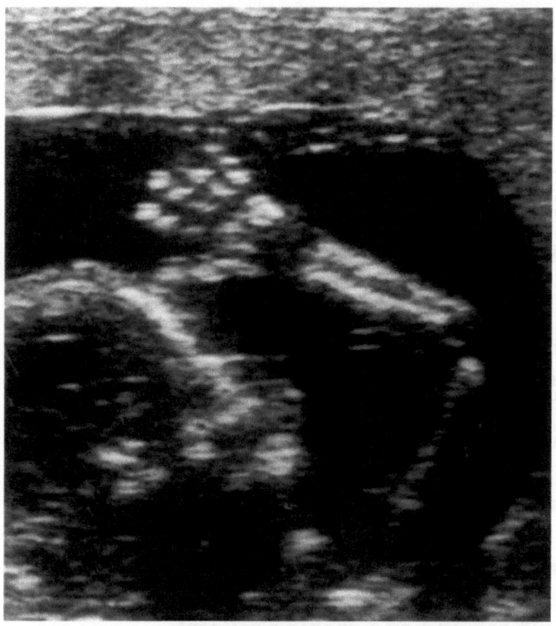

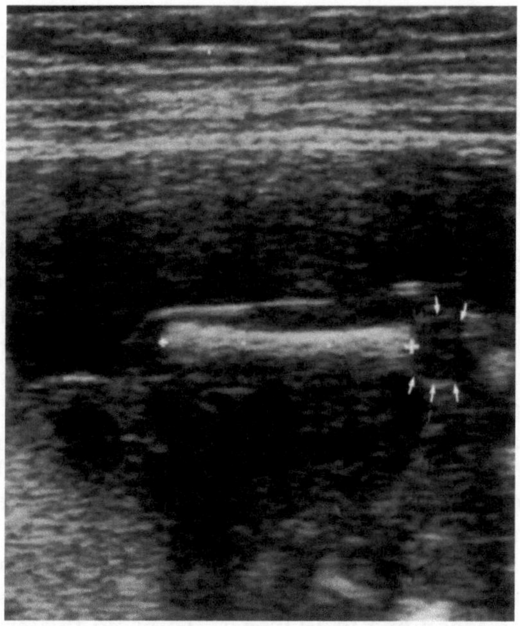

Fig. 10.15. (left) Scan of the upper extremity at 17 weeks, displaying the individual phalanges, metacarpals, ulna, radius, and humerus all in one scan

Fig. 10.16. (right) Measurement of humeral length in an 18-week fetus. The unossified humeral head *(arrows)* contrasts with the surrounding soft tissues

positioned. The humerus is shown in Fig. 10.16. The cartilaginous structures of the humeral head are visible on the right side of the image, along with portions of the glenoid roof, which is not yet ossified but already appears denser than the head. The forearm bones can be individually identified from the anatomic orientation of the scan plane and from the slightly greater length of the ulna in relation to the radius (Fig. 10.17). Hansmann et al. (1985) have published comprehensive tables of normal values for these bones. The fingers can be individually defined by ultrasound from 12–13 weeks (Figs. 10.18 and 10.19), but this is too time-consuming for routine studies, because counting and identifying the individual fingers at this early stage would necessitate imaging the extended hand on a single plane.

Our observations indicate that the motor activity of the fetal hand is substantial. Alternating episodes of hand opening and fist clenching are frequently observed in the alert state. It is rare to demonstrate the fingers at a time when they are simultaneously spread apart and extended (Fig. 10.20). It is most common to find the fingers closed in fist-like fashion (Fig. 10.21), with the thumb held against the palm and rarely occupying the same plane as the other digits. The importance of diagnosing deformities of the fingers was stated previously (Staudach and Lassmann 1984; Hansmann et al. 1985; Jeanty et al. 1985b). One should be particularly alert for permanent flexion of the metacarpophalangeal joints, the permanent crossing of individual fingers, and abnormalities of number.

10.6.4 Lower Extremity

The lower extremities are considerably easier to image than the upper extremities, because leg and feet movements are of lesser amplitude and occur less often. It is common for the legs to be held in a crossed position, with a corresponding superimposition of bony structures, and it is essential that the knee and elbow joints be used as pivot points for rotating the transducer into the desired image plane (Fig. 10.25 a, b). Figure 10.26 shows a typical lower extremity posture in a 13-week fetus. The ossified segments of the femora are visible in the upper part of the limbs, and the legs are crossed below the knees.

Femur length measurement. The femur can be demonstrated and measured on transverse as well as sagittal scans. It is important that the femur located closer to the transducer be used for the length measurement. On transverse scans, the femur located farther from the transducer may be partially obscured by the acoustic shadow from the opposite leg (Fig. 10.30). On sagittal scans the femur, knee joint, tibia, and fibula can all be demonstrated concurrently (Figs. 10.27–10.29). It should be kept in mind that the true length of a bone is measurable only if the scan plane traverses the longest axis of that bone. Figures 10.28 and 10.29 illustrate the discrepancy that can arise from an incorrect measurement. The scan plane in Fig. 10.28 traverses the femur at a slightly oblique angle, and the measured length, at 20 mm, corresponds to only about 16 weeks' gestation (Hansmann et al. 1985). Rotating the transducer to visualize the total femur length yields a measurement of 31 mm, corresponding to 19 weeks. To avoid this error, we recommend making several femur length measurements and then taking the longest measurement for correlation with gestational age. The echo intensity of the femur should be uniform over its entire length, and the femur should cast a uniform acoustic shadow. On a sagittal section through the knee joint (Fig. 10.31), the condyles of the femur and tibia

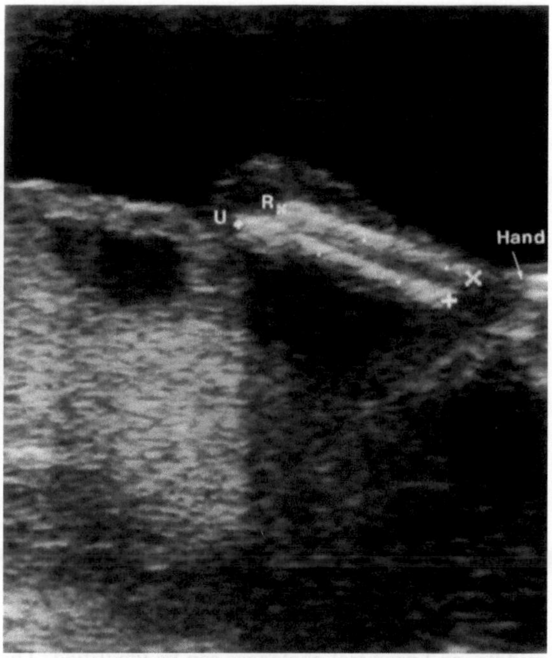

Fig. 10.17. Measurement of the length of the ulna (*U*) and radius (*R*) in an 18-week fetus. The fingers (not shown) are at the right edge of the image

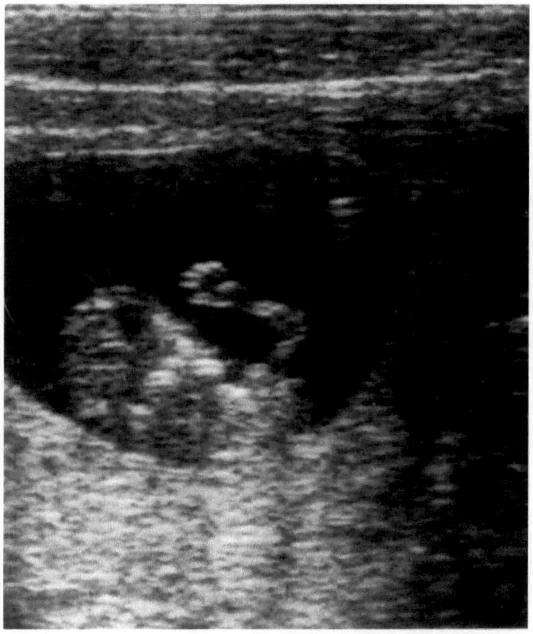

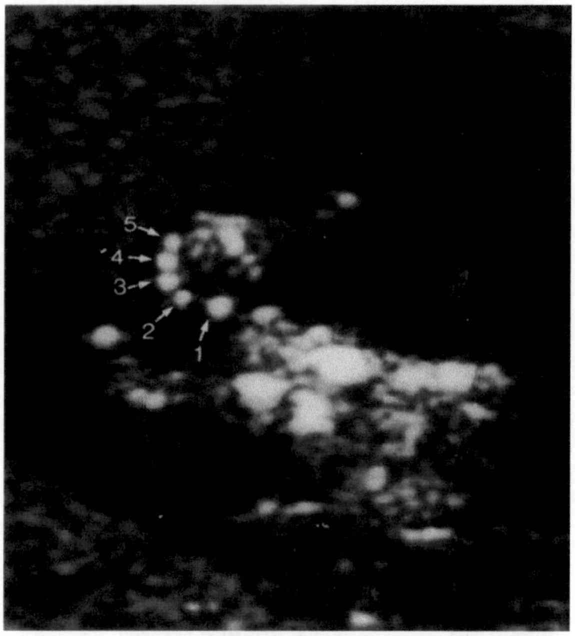

Fig. 10.18. (left) Earliest detection of the fingers in a 12-week fetus

Fig. 10.19. (right) Fetal fingers at 12 weeks. It is possible to count the fingers on the close-up view

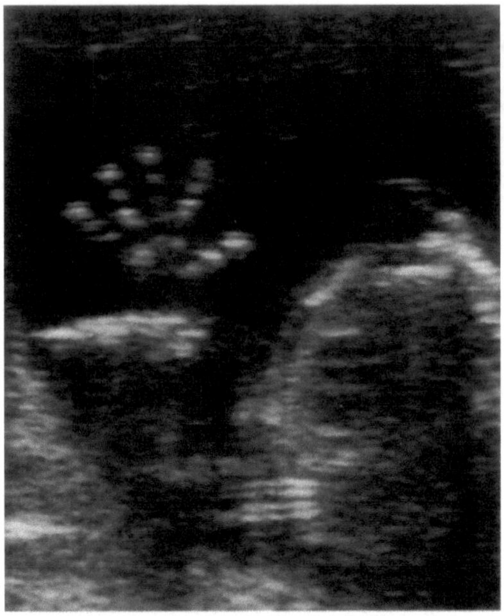

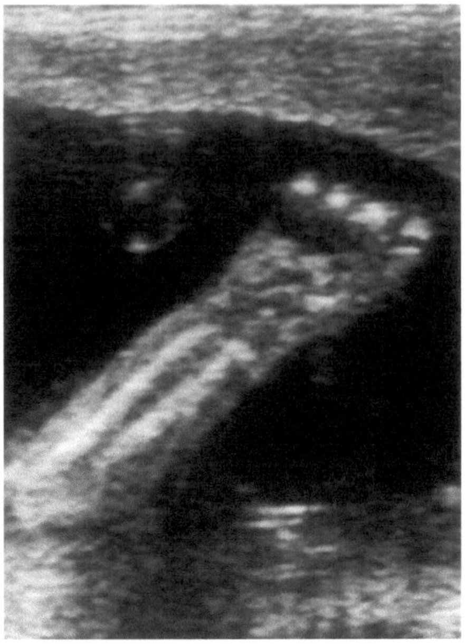

Fig. 10.20. (left) Simultaneous visualization of all five fingers of a 16-week fetus. The distal, middle, and proximal phalanges can be identified. The scan demonstrates the extended digits in one plane

Fig. 10.21. (right) Here the hand is closed into a fist, and the scan plane must be adjusted slightly to visualize the thumb

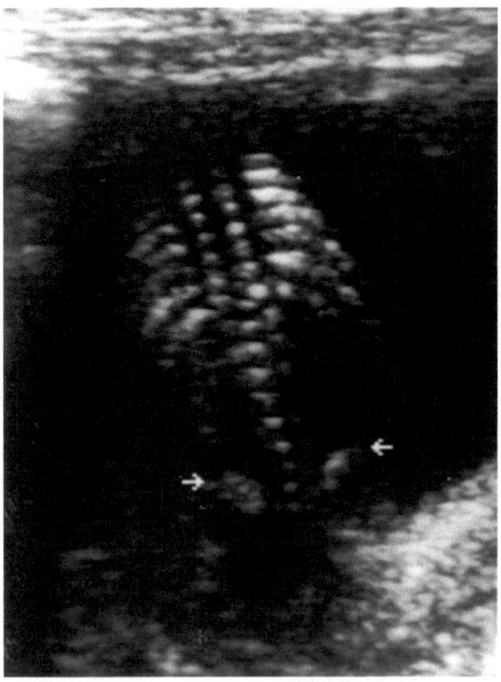

Fig. 10.22. Frontal scan through a fetus at 13 weeks. Cranially to caudally we observe the ribs, the ossification centers of the lumbar vertebral bodies, and the ossification centers of the ilia *(arrows)*

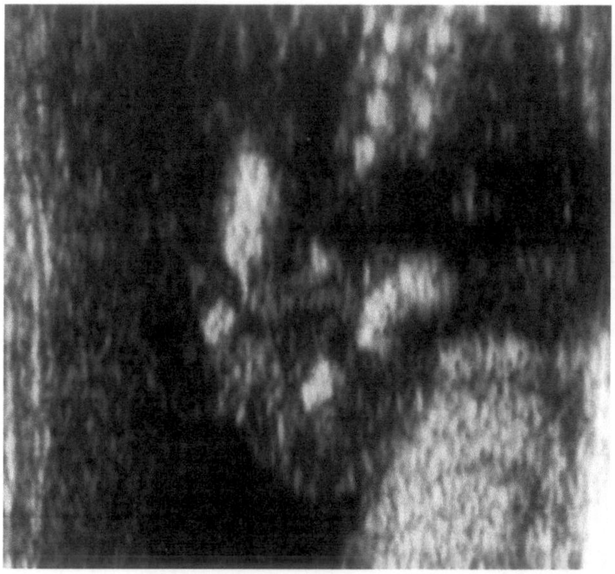

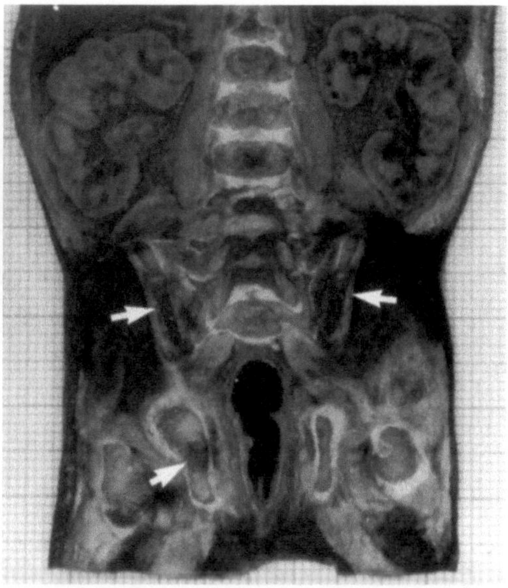

Fig. 10.23. a Frontal scan of the fetal pelvis at 18 weeks (image rotated 90°). The prominent ossification centers of the ilium and ischium are seen caudal to the lumbar spine

Fig. 10.23. b Analogous frozen section at 21 weeks. The ossification centers of the ilia *(top arrows)* appear darker than their surroundings. This section contains the ossification center of the left ischium *(bottom arrow)*, but not of the right

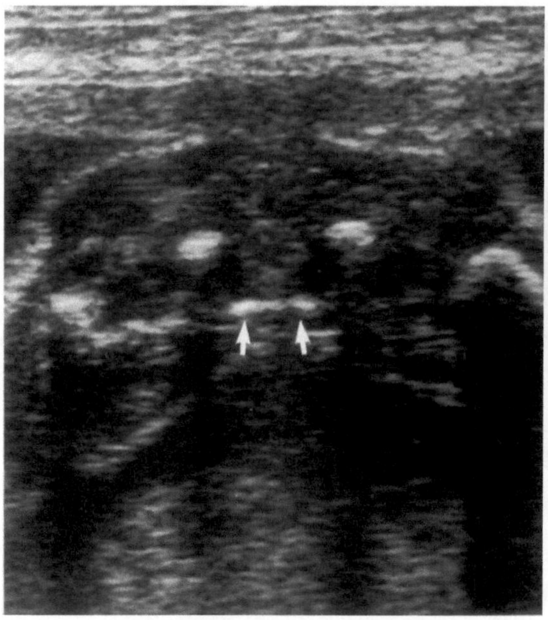

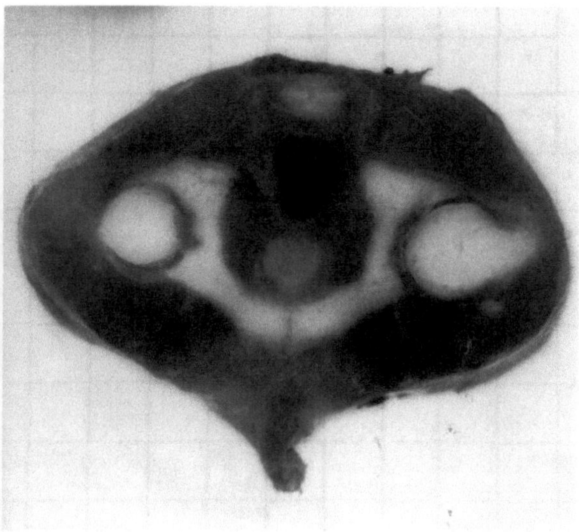

Fig. 10.24 a, b. PA transverse scan of the pelvis at 23 weeks showing the pubic ossification centers *(arrows)*, located below and anterior to the ossification centers of the ischium

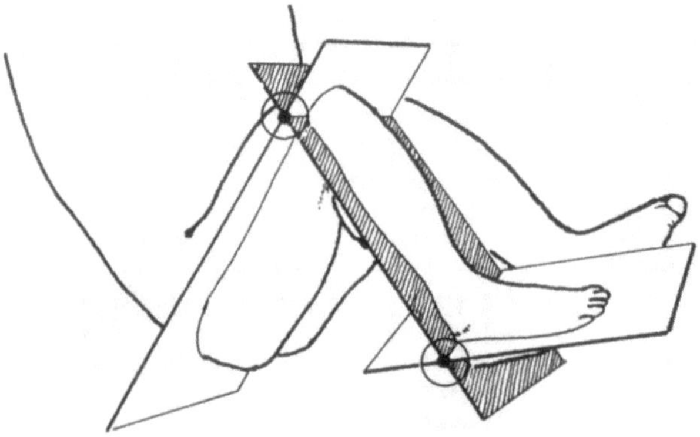

Fig. 10.25. a Diagram showing the spatial orientation of the lower extremity image planes

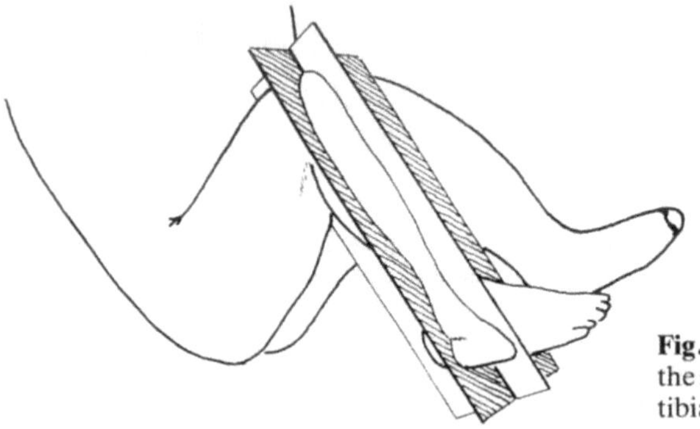

Fig. 10.25. b Diagram showing how the scan plane is rotated to image the tibia and fibula

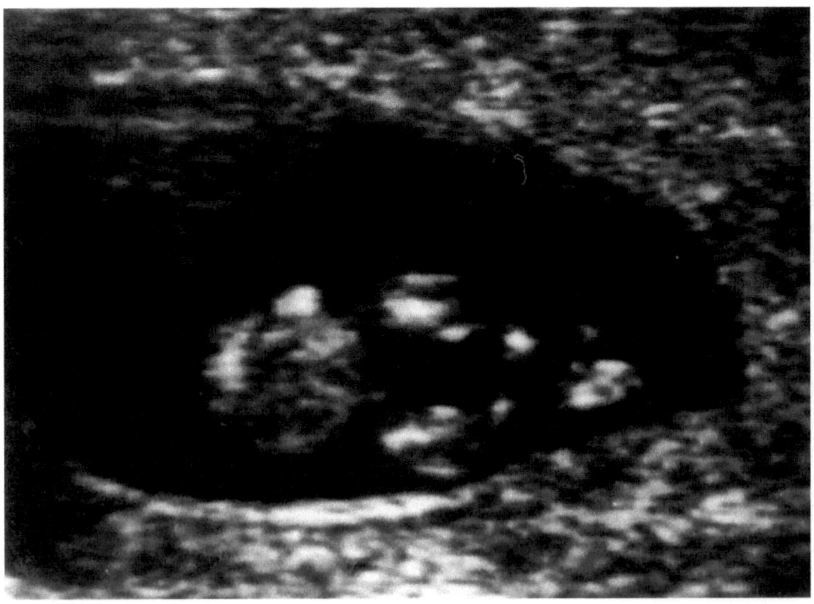

Fig. 10.26. Scan through the lower extremities of a 12-week fetus. The lower legs are crossed, and the ossified portions of the femoral shafts are visible

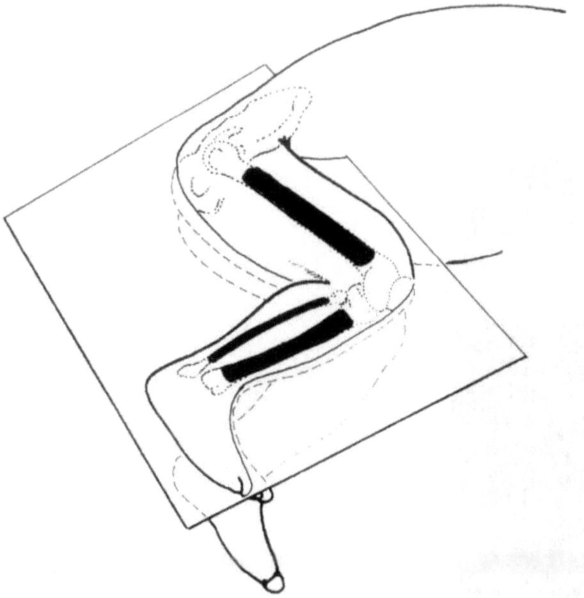
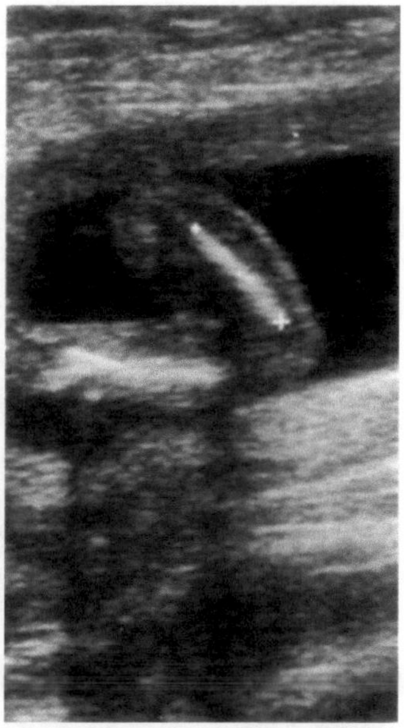

Fig. 10.27. (left) Schematic representation of the sagittal image plane through the lower extremity

Fig. 10.28. (right) Sagittal scan of the lower extremity at 19 weeks. This scan does not demonstrate the full longitudinal extent of the femur, so the length measurement (only 20 mm) is inaccurate

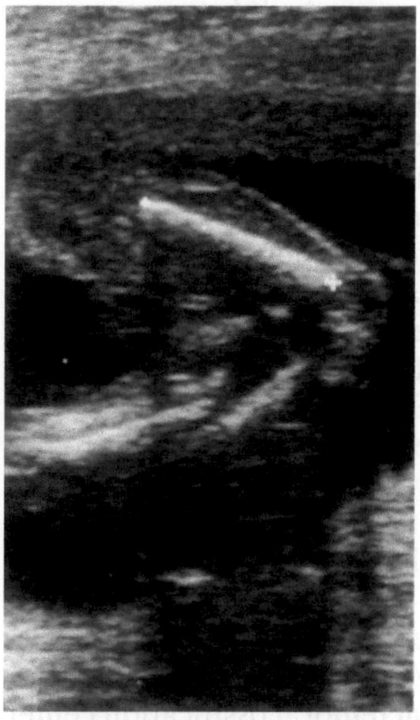
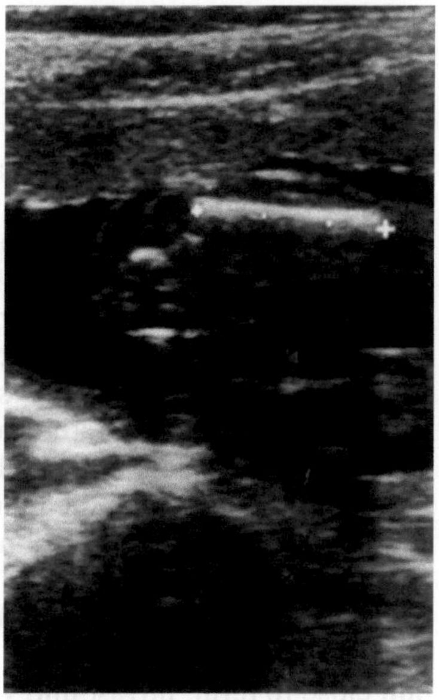

Fig. 10.29. (left) Sagittal scan of the same extremity at 19 weeks. Here an accurate femur length measurement (31 mm) is obtained

Fig. 10.30. (right) Femur length measurement in a 19-week fetus on a frontal scan through the thigh. The femur close to the transducer is measured; the more distant femur is obscured by acoustic shadowing

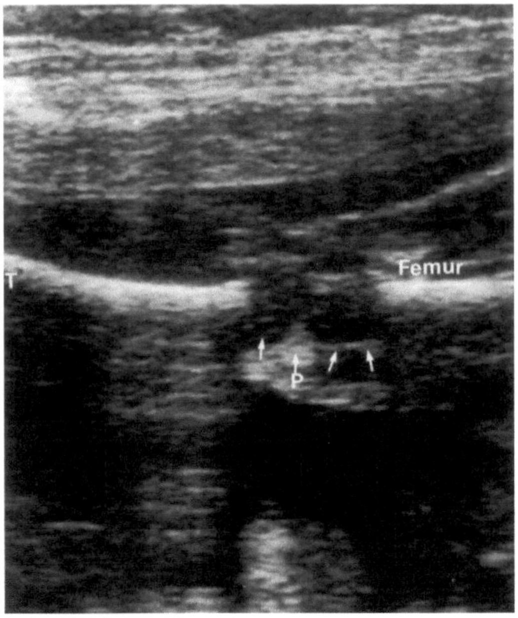

Fig. 10.31. Sagittal scan through the knee joint of a 27-week fetus. The femur is on the right, the tibia on the left. The patella (*P*) appears below the articulating condyles

appear as spherical structures whose echogenicity contrasts subtly but recognizably with surrounding tissues. Between the condyles, the developing patella is visible anteriorly as a bright echo.

Increasing attention is given to the axial alignment of the foot relative to the lower leg as a means of diagnosing deformities in that region. The prenatal detection of clubfeet has been reported by Staudach et al. (1984), Hansmann et al. (1985), Chervenak et al. (1985), and Jeanty et al. (1985b). This diagnosis relies on accurate spatial orientation with regard to the limb axes. Normally a sagittal scan through the foot will show its typical angulation relative to the lower leg and the contours of the heel and sole (Figs. 10.32 a, b). On a frontal scan through the lower leg, the foot itself normally will not appear angulated relative to the axis of the lower leg. If the sole of the foot is visible on a scan depicting both the tibia and fibula, an abnormal angular deformity should be assumed (Fig. 10.33 a, b). When such deformities are detected, the patient should be referred for exclusion of syndromes and chromosomal disorders. The toes are most easily visualized and counted on a tangential scan through the plantar surface of the foot. Generally it is far easier to count the toes than the fingers owing their relatively stationary position (Fig. 10.34).

Bernaschek (1982) has stressed the importance of identifying the ossification centers of the lower extremity and assessing their size. When gestational age is uncertain close to term, ultrasound visualization of the distal femoral ossification center and proximal tibial ossification center can provide a rough estimate (Fig. 10.35), although we have found that the wide individual variation in the appearance of these centers precludes an accurate assessment of gestational age.

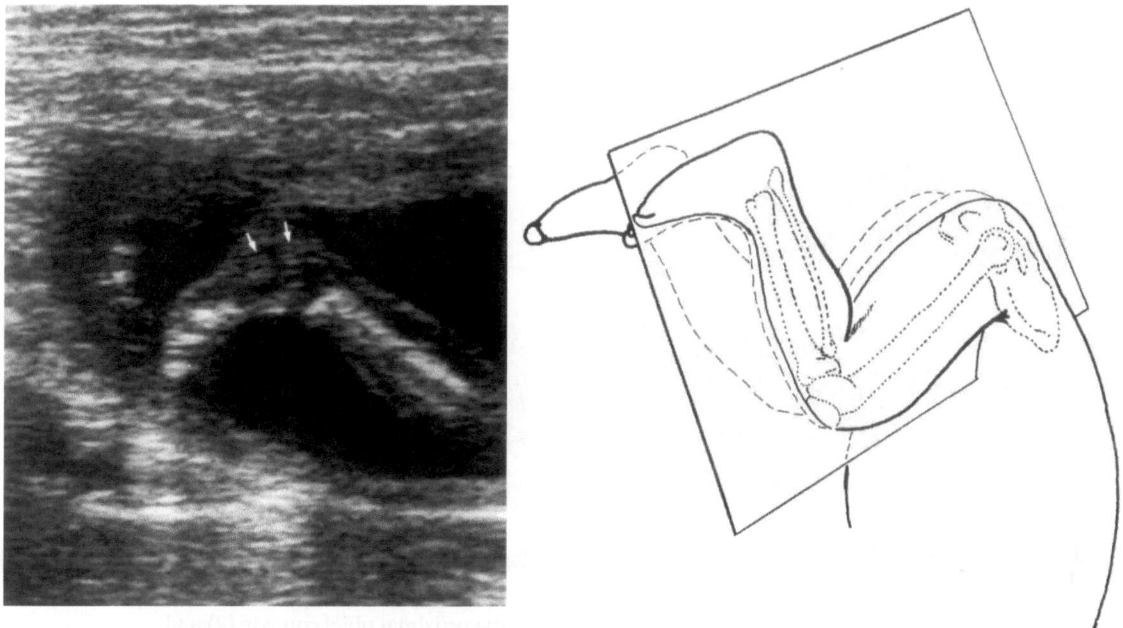

Fig. 10.32. a Sagittal scan through the lower leg of a 17-week fetus. The position of the foot is normal in relation to the lower leg, and the densities of the talus and calcaneus can be recognized even before the appearance of true ossification centers

Fig. 10.32. b Schematic drawing of the image plane

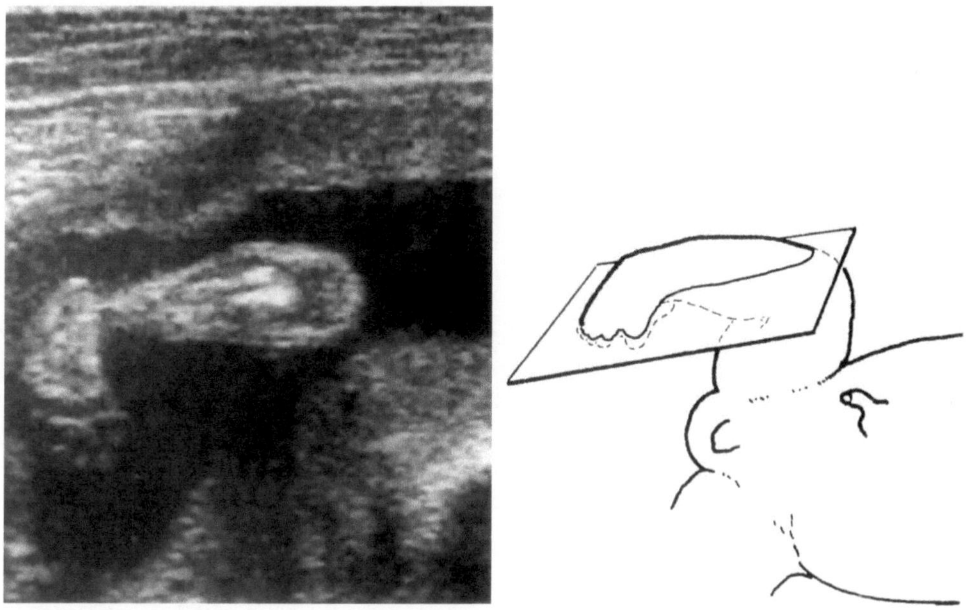

Fig. 10.33. a Sonogram showing a clubfoot deformity at 24 weeks. The long axis of the lower leg and the entire sole of the foot are depicted in one plane

Fig. 10.33. b Schematic drawing of the image plane

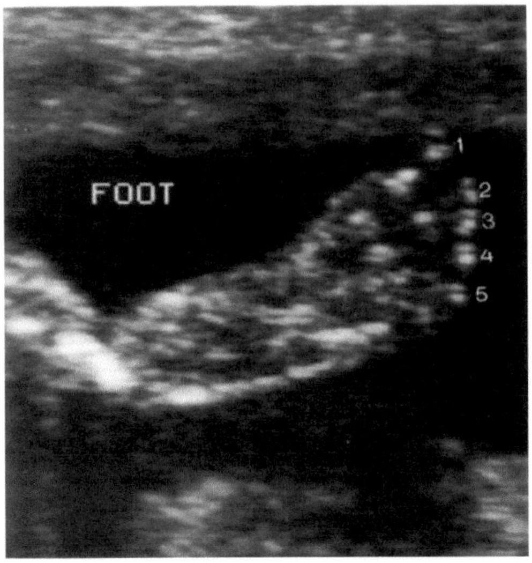

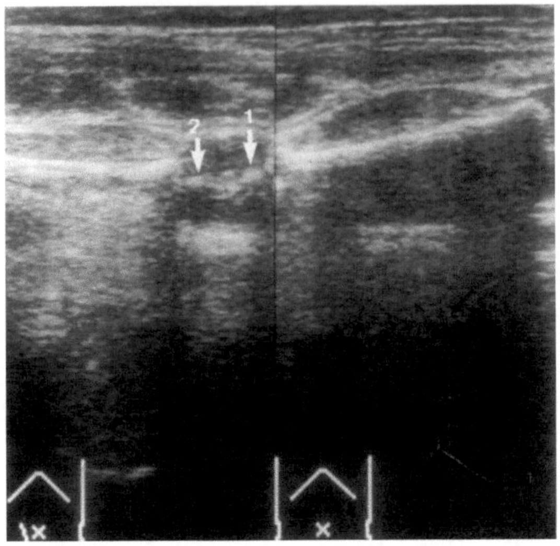

Fig. 10.34. (left) Visualization of the foot on a scan through the plantar surface. The toes are easily identified and counted

Fig. 10.35. (right) Demonstration of the ossification centers of the proximal tibial condyle (*2*) and distal femoral condyle (*1*)

References

Chapter 1

Campbell S, Thoms A (1982) Ultrasound measurement of fetal head to abdomen circumference ratio in the assessment of growth retardation. Br J Obstet Gynaecol 89:165

Chinn DH, Filly RA, Callen PW (1982) Ultrasonic evaluation of fetal umbilical and hepatic vascular anatomy. Radiology 144:153

Grant EG, Schellinger D, Borts FT (1981) Realtime sonography of the neonatal and infant head. Am J Roentgenol 136:265

Hadlock FP, Deter RL, Park SK (1981) Realtime sonography: ventricular and vascular anatomy of the fetal brain in utero. Am J Roentgenol 136:133

Jeanty P, Chervenak FA, Romero R (1984) The sylvian fissure: a commonly mislabeled cranial landmark. J Ultrasound Med 3:15

Johnson ML, Rumack CM (1980) Ultrasonic evaluation of the neonatal brain. Radiol Clin North Am 18:117

Chapter 2

Bahr FG, Bloom G, Friberg U (1957) Volume changes of tissue in physiological fluids during fixation in osmium tetroxide or formaldehyde and during subsequent treatments. Exp Cell Res 12:342

Bernaschek G, Dadak C, Kratochwil A (1980) Frühzeitige Diagnose fetaler Mißbildungen. Geburtshilfe Frauenheilkd 40:868

Boonstra H, Oosterhis JW, Oosterhuj AM, Fleuren GJ (1984) Cervical tissue shrinkage by formaldehyde fixation, paraffin wax embedding, section cutting and mounting. Virchows Archiv A 402:195

Födisch HJ (1982) Pathologisch-anatomische Mißbildungsdiagnostik – Heute. Verh Dtsch Ges Pathol 66:37

Födisch HJ, Knöpfle G (1984) Patho-anatomische Teratologie – eine aktuelle Herausforderung. Gynäkologe 17:2

Grannum P, Bracken M, Silverman R, Hobbins JC (1980) Assessment of fetal kidney size in normal gestation by comparison of ratio of kidney circumference to abdominal circumference. Am J Obstet Gynecol 136:249

Hansmann M (1981) Nachweis und Ausschluß fetaler Entwicklungsstörungen mittels Ultraschallscreening und gezielter Untersuchung – ein Mehrstufenkonzept. Ultraschall 2:206

Hansmann M, Gembruch U (1984) Gezielte sonographische Ausschlußdiagnostik fetaler Fehlbildungen in Risikogruppen. Gynäkologe 17:19

Hansmann M, Hackelöer BJ, Staudach A (1985) Ultraschalldiagnostik in Geburtshilfe und Gynäkologie. Springer, Berlin Heidelberg New York Tokyo

Hobbins JC, Grannum PAT, Berkowitz RL, Silverman R, Mahoney MJ (1979) Ultrasound in the diagnosis of congenital anomalies. Am J Obstet Gynecol 134:331

Klemstein J (1981) Die Entwässerung voluminöser Organe zur Plastination unter Vermeidung von starken Schrumpfungen. Präparator 27:169

Kushida H (1962) A study of cellular swelling and shrinkage during fixation, dehydration and embedding in various standard media. J Electronmicrosc 11:135

Rehder H (1982) Fetalpathologie im Rahmen pränataler Diagnostik. Verh Dtsch Ges Pathol 66:58

Staudach A (1982) Möglichkeiten und Grenzen der Mißbildungsdiagnostik. Swiss Med 4:67
Staudach A, Laßmann R, Rosenkranz W, Engels M, Joos H, Rücker J (1984) Praenatale Diagnose fetaler Entwicklungsstörungen – das Modell eines interdisziplinären Teams. In: Kowalewski S (Hrsg) Pädiatrische Intensivmedizin VI, Thieme, Stuttgart New York
Tsukasa J, Mori H, Ishiguro K, Takeishi M (1984) Dimensional changes of tissues in the course of processing. J Microsc 136:323
Weiß H, Zerres K, Hansmann M (1981) Pränatale Diagnose zystischer Nierenveränderungen mit Hilfe der Ultraschalltechnik. Ultraschall 2:205
Winter R (1981) Die Diagnose angeborener Mißbildungen mittels Ultraschall. Ultraschall 2:235

Chapter 3

Chamberlain PF, Manning FA, Morrison I, Harman CR, Lange IR (1984a) I. The relationship of marginal and decreased amniotic fluid volumes to perinatal outcome. Am J Obstet Gynecol 150:245
Chamberlain PF, Manning FA, Morrison I, Harman CR, Lange IR (1984b) II. The relationship of increased amniotic fluid volume to perinatal outcome. Am J Obstet Gynecol 150:250
Crowley P, O'Herlihy C, Boylan P (1984) The value of ultrasound measurement of amniotic fluid volume in the management of prolonged pregnancies. Br J Obstet Gynaecol 91:444
Göttlicher S, Madjarić J, Krone HA (1981) Über die Lage des menschlichen Feten und die Wahrscheinlichkeit einer spontanen Lageveränderung im Verlauf der Schwangerschaft. Z Geburtshilfe Perinatol 185:288
Halperin ME, Fong KW, Zalev AH, Goldsmith CH (1985) Reliability of amniotic fluid volume estimation from ultrasonograms: Intraobserver and interobserver variation before and after the establishment of criteria. Am J Obstet Gynecol 153:264
Hansmann M, Hackelöer BJ, Staudach A (1985) Ultraschalldiagnostik in Geburtshilfe und Gynäkologie. Springer, Berlin Heidelberg New York Tokyo
Hill LM, Breckle R, Wolfgram KR, O'Brien PC (1983) Oligohydramnios: ultrasonically detected incidence and subsequent fetal outcome. Am J Obstet Gynecol 47:407
Klug PW, Staudach A, Hohlweg T (1985) Cervixsonographie versus Palpation. In: Otto R, Schaars P (Hrsg) Ultraschalldiagnostik 1985. Thieme, Stuttgart New York
Manning FA, Hill LM, Platt LD (1981) Qualitative amniotic fluid volume determination by ultrasound: antepartum detection of intrauterine growth retardation. Am J Obstet Gynecol 139:254
Philipson EH, Sokol RI, Williams T (1983) Oligohydramnios: clinical associations and predictive value for intrauterine growth retardation. Am J Obstet Gynecol 146:271
Queenan JT, Thompson W (1972) Amniotic fluid volumes in normal pregnancies. Am J Obstet Gynecol 114:34
Reading AE, Cox DN (1982) The effects of ultrasound examination on maternal anxiety levels. J Behav Med 5/2:237
Staudach A (1984) Möglichkeiten und Grenzen der geburtshilflichen Endoskopie. Der praktische Arzt 38:1897

Chapter 4

Babcock DS, Han BK (1981) Cranial Ultrasonography of infants. Williams & Wilkins, Baltimore London
Birnholz JC (1981) The development of human fetal eye movement patterns. Science 213:679
Birnholz JC (1982) Newborn cerebellar size. Pediatrics 70:284
Birnholz JC (1983) Fetal behavior and condition. In: Callen PW (ed) Ultrasonography in obstetrics and gynecology. Saunders, Philadelphia
Birnholz JC (im Druck) Fetal and infant brain development. In: Otto J, Schnaars P (Hrsg) Ultraschalldiagnostik 1985 – Drei-Ländertreffen Zürich. Thieme, Stuttgart New York
Bots RS, Nijhuis JG, Martin CB, Prechtl HFR (1981) Human fetal eye movements: detection in utero by ultrasonography. Early Hum Dev 5:87
Campbell S (1968) An improved method of fetal cephalometry by ultrasound. Br J Obstet Gynaecol 75:568

Campbell S (1979) Early prenatal diagnosis of fetal abnormality by ultrasound B-scanning. In: Prenatal Diagnosis. Enke, Stuttgart

Campbell S, Pearce JM (1983) The prenatal diagnosis of fetal structural anomalies by ultrasound. Clin Obstet Gynecol 10/3:475

Crade M, Patel J, McQuown D (1981) Sonographic imaging of the glycogen stage of the fetal choroid plexus. Am J Neurol Rad 2:345

Davies DV, Davies F (1962) Gray's anatomy, ed 33. Longmans, Green & Co Ltd, London

Denkhaus H, Winsberg F (1979) Ultrasonic measurements of the fetal ventricular system. Radiology 131:781

Donald I, Brown TG (1961) Demonstration of tissue interfaces within the body by ultrasonic echosounding. Br J Radiol 34:539

Dorovini-Zis K, Dolman CL (1977) Gestational development of brain. Arch Pathol Lab Med 101:192

Fiske CE, Filly RA, Callen PW (1981) Sonographic measurement of lateral ventricular width in early ventricular dilation. J Clin Ultrasound 9:303

Grant EG, Schellinger D, Borts FT (1981) Realtime sonography of the neonatal and infant head. Am J Roentgenol 136:265

Hadlock FP, Deter RL, Park SK (1981 a) Realtime sonography: ventricle and vascular anatomy of the fetal brain in utero. Am J Roentgenol 136:133

Hadlock FP, Deter RL, Carpenter RJ, Park SK (1981b) Estimating fetal age: effect of head shape on BPD. Am J Roentgenol 137:83

Hansmann M (1976) Ultraschallbiometrie im II. und III. Trimester in der Schwangerschaft. Gynäkologe 9:133

Hansmann M (1981) Nachweis und Ausschluß fetaler Entwicklungsstörungen mittels Ultraschallscreening und gezielter Untersuchung – ein Mehrstufenkonzept. Ultraschall 2:206

Hansmann M, Hackelöer BJ, Staudach A (1985) Ultraschalldiagnostik in Geburtshilfe und Gynäkologie. Springer, Berlin Heidelberg New York Tokyo

Hobbins JC, Grannum PAT, Berkowitz RL, Silverman R, Mahoney MJ (1979) Ultrasound in the diagnosis of congenital anomalies. Am J Obstet Gynecol 134:331

Hobbins JC, Winsberg F, Berkowitz RL (1983) Ultrasonography in obstetrics and gynecology. Williams & Wilkins, Baltimore London

Hofmann D, Holländer HJ (1968) Über den Nachweis fetalen Lebens und die Messung des kindlichen Schädels mittels des zweidimensionalen Ultraschallechoverfahrens. Gynaecologia 165:60

Holländer HJ (1972, 1975, 1984) Die Ultraschalldiagnostik in der Schwangerschaft. Urban & Schwarzenberg, München Berlin Wien

Hopf HC, Poeck K, Schliak H (1984) Neurologie in Praxis und Klinik. Georg Thieme, Stuttgart New York

Jeanty P, Cantraine F, Cousaert E, Romero R, Hobbins JC (1984a) The binocular distance: a new way to estimate fetal age. J Ultrasound Med 3:241

Jeanty P, Chervenak FA, Romero R, Michiels M, Hobbins JC (1984b) The sylvian fissure: a commonly mislabeled cranial landmark. J Ultrasound Med 3:15

Johnson ML, Dunne MG, Mack LA, Rashbaum CM (1980a) Evaluation of fetal intracranial anatomy by static and real-time ultrasound. J Clin Ultrasound 8:311

Johnson ML, Rumack CM (1980b) Ultrasonic evaluation of the neonatal brain. Radiol Clin North Am 18:117

Kasby CB, Poll V (1982) The breech head and its ultrasound significance. Br J Obstet Gynaecol 89:106

Kier EL (1971) Fetal skull. In: Radiology of the skull and brain. Mosby, USA (Volume I, Book I, p 99)

Kier EL (1974) Fetal cerebral arteries: a phylogenetic and ontogenetic study. In: Radiology of the skull and brain. Mosby, USA (Volume II, Book I. p 1089)

Kier EL (1977) The cerebral ventricles: a phylogenetic and ontogenetic study. In: Radiology of the skull and brain. Mosby, USA (Volume III, p 2787)

Kurtzke JF, Goldberg ID, Kurland LT (1973) The distribution of deaths from congenital malformations of the nervous system. Neurology (Minneap) 23:483

Levi S, Erbsman F (1975) Antenatal fetal growth from the nineteenth week. Am J Obstet Gynecol 121:262

Mahony BS, Callen P, Filly R, Hoddick K (1984) The fetal cisterna magna. Radiology 153:773

Mayden K, Tortora M, Berkowitz RL, (1982) Orbital diameters: a new parameter for prenatal diagnosis and dating. Am J Obstet Gynecol 144:289

McFarland WL, Morgane PJ, Jacobs MS (1969) Ventricular system of the brain of the dolphin, Tursiops truncatus, with comparative anatomical observations and relations to brain specializations. J Comp Neurol 135:275
McGahan JP, Phillips HE, Ellis WG (1983) The fetal hippocampus. Radiology 147:201
McLeary RD, Kuhns LR, Barr J (1984) Ultrasonography of the fetal cerebellum Radiology 151:439
Nijhuis JG, Prechtl HFR, Martin CB, Bots RS (1982) Are there behavioural states in the human fetus? Early Hum Dev 6:177
Perry RNW, Bowman ED, Murton LJ, Roy RND, DeCrespigny LC (1985) Ventricular size in newborn infants. J Ultrasound Med 4:475
Schillinger H, Müller R, Kretzschmar M, Wode J (1976) Bestimmung des Gestationsalters in der Spätschwangerschaft durch Ultraschall. Geburtsh Frauenheilkd 36:500
Schmid F (1973) Pädiatrische Radiologie. Band I. Springer, Berlin Heidelberg New York
Smith DW (1982) Recognizable patterns of human malformations. Saunders, Philadelphia London Toronto
Staudach A, Laßmann R (1984) Ultraschalldiagnostik von fetalen Mißbildungen. Oester Aerztetg 39/7:476
Vintzileos AM, Ingardia CJ, Nochimsen DJ (1983) Congenital hydrocephalus: a review and protocol for perinatal management. Obstet Gynecol 62:539
Weisberg LA, Nice C, Katz M (1978) Cerebral computed tomography: a text-atlas. Saunders, Philadelphia London Toronto
Willocks J (1963) Fetal cephalometry by ultrasound. Thesis, Glasgow
Willocks J, Donald I, Duggan IC, Day N (1964) Foetal cephalometry by ultrasound. Br J Obstet Gynaecol 71:11
Winter R (1981) Die Diagnose angeborener fetaler Mißbildungen mittels Ultraschall. Ultraschall 2:235

Chapter 5

Hansmann M, Gembruch U (1984) Gezielte Ausschlußdiagnostik fetaler Entwicklungsstörungen. Gynäkologe 17:19
Hansmann M, Hackelöer BJ, Staudach A (1985) Ultraschalldiagnostik in Geburtshilfe und Gynäkologie. Springer, Berlin Heidelberg New York Tokyo
Leucht W, Müller E, Heyes H, Töllner U, Jonatha W (1979) Probleme bei der pränatalen Diagnose von Neuralrohrdefekten. Z Geburtshilfe Perinatol 183:434
Miskin M, Baim RS, Allen LC, Benzie RJ (1979) Ultrasonic assessment of the fetal spine before 20 weeks' gestation. Radiology 132:131
Rickham PP, Soper RT, Stauffer UG (1975) Kinderchirurgie. Thieme, Stuttgart New York

Chapter 6

Bowie JD, Clair MR (1982) Fetal swallowing and regurgitation: observation of normal and abnormal activity. Radiology 144:877
Cooper C, Mahony BS, Bowie JD, Albright TO, Callen PW (1985) Ultrasound evaluation of the normal fetal upper airway and esophagus. J Ultrasound Med 4:343
Eyheremendy E, Pfister M (1983) Antenatal real-time diagnosis of esophageal atresias. J Clin Ultrasound 11:395
Farrant P (1980) The antenatal diagnosis of oesophageal atresia by ultrasound Br J Radiol 53:1202
Jeanty P, Romero R, Hobbins JC (1984) Vascular anatomy of the fetus. J Ultrasound Med 3:113
Pretorius DH, Meier PR, Johnson ML (1983) Diagnosis of esophageal atresia in utero. J Ultrasound Med 2:475
Utsu M, Sakakibara S, Ishida T (1983) Dynamics of tracheal fluid flow in the human fetus, studied with pulsed Doppler ultrasound. Acta Obstet Gynecol Jpn 35:2017

Chapter 7

Allan LD, Tynan MJ, Campbell S, Wilkinson JL, Anderson RH (1980) Echocardiographic and anatomical correlates in the fetus. Br Heart J 44:444

Davis CL (1982) Diagnosis and management of nonimmune hydrops fetalis. J Reprod Med 27:594

DeVore GR, Donnerstein RL, Kleinman CS, Platt LD, Hobbins JC (1982) Fetal echocardiography. I. Normal anatomy as determinded by real-time-directed M-mode ultrasound. Am J Obstet Gynecol 144:249

DeVore GR, Siassi B, Platt LD (1983) Fetal echocardiography. III. The diagnosis of cardiac arrhythmias using real-time-directed M-mode ultrasound. Am J Obstet Gynecol 146:792

DeVore GR, Siassi B, Platt LD (1984) Fetal echocardiography. IV: M-mode assessment of ventricular size and contractility during the second and third trimesters of pregnany in the normal fetus. Am J Obstet Gynecol 150:981

DeVore GR, Platt LD (1985) The random measurement of the transverse diameter of the fetal heart: a potential source of error. J Ultrasound Med 4:335

DeVore GR, Siassi B, Platt LD (1985) The use of the abdominal circumference as a means of assessing M-mode ventricular dimensions during the second and third trimesters of pregnancy in the normal human fetus. J Ultrasound Med 4:175

Garrett WJ, Robinson DE (1970) Fetal heart size measured in vivo by ultrasound. Pediatrics 46:1

Garrett WJ (1979) Ultrasound in discerning normal fetal anatomy. In: Hobbins JC (ed) Diagnostic ultrasound in obstetrics. Churchill Livingstone, New York Edinburgh London

Grube E (1985) Zweidimensionale Echocardiographie. Thieme, Stuttgart New York

Hansmann M, Redel DA (1982) Prenatal symptoms and clinical management of heart disease. In: 1er symposium international d'echocardiologie foetale, Strasbourg 1982, p 137

Hansmann M, Redel DA, Födisch HJ (1982) Premature obstruction of the foramen ovale detected, treated and reconfirmed by help of ultrasound. In: Burruto F, Hansmann M, Wladimiroff JW (eds) Fetal ultrasonography: The secret prenatal life. Wiley, Chichester New York, p 151

Hansmann M, Gembruch U (1984) Gezielte sonographische Ausschlußdiagnostik fetaler Fehlbildungen in Risikogruppen. Gynäkologe 17:19

Hansmann M, Hackelöer BJ, Staudach A (1985) Ultraschalldiagnostik in Geburtshilfe und Gynäkologie. Springer, Berlin Heidelberg New York Tokyo

Jeanty P, Romero R, Cantraine F, Cousaert E, Hobbins JC (1984a) Fetal cardiac dimensions: a potential tool for the diagnosis of congenital heart defects. J Ultrasound Med 3:359

Jeanty P, Romero R, Hobbins JC (1984b) Fetal pericardial fluid: a normal finding of the second half of gestation. Am J Obstet Gynecol 149:529

Jeffrey RB, Laing FC (1982) High-resolution real-time sonography of fetal cardiovascular anatomy. J Ultrasound Med 1:249

Kleinman CS, Hobbins JC, Jaffe CC, Lynch DC, Talner NS (1980) Echocardiographic studies of the human fetus: prenatal diagnosis of congenital heart disease and cardiac dysrhythmias. Pediatrics 65:1059

Kleinman CS, Donnerstein RL, DeVore GR et al (1982a) Fetal echocardiography for evaluation of in utero congestive heart failure. N Engl J Med 306:568

Kleinman CS, Talner NS, Donnerstein RL, DeVore GR, Hobbins JC (1982b) Fetal echocardiography for evaluation of in utero cardiac dysrhythmias. In: 1er symposium international d'echocardiolgie foetale, Strasbourg, p 217

Kleinman CS, Donnerstein RL, Jaffe CC et al (1983) Fetal echocardiography. A tool for evaluation of in utero cardiac arrhythmias and monitoring of in utero therapy: Analysis of 71 patients. Am J Cardiol 51:237

Köhler C, Schumacher G, Meierhofer JN, Peter B (1981) Pränatale Ultraschalldiagnostik eines schweren Herzvitiums. Geburtshilfe Frauenheilkd 41:36

Leslie J, Shen S, Thornton JC, Strauss L (1983) The human fetal heart in the second trimester of gestation: a gross morphometric study of normal fetuses. Am J Obstet Gynecol 145:312

Levi S, Erbsman F (1975) Antenatal fetal growth from the nineteenth week. Am J Obstet Gynecol 121:262

Nisand I, Spielmann A, Dellenbach P (1984) Fetal heart-present investigative means. Ultrasound in Med Biol 10:79

Redel DA, Hansmann M (1981) Fetal obstruction of the foramen ovale detected by two-dimensional Doppler echocardiography. In: Rijsterborgh H (ed) Echocardiology. Nijhoff, The Hague Boston London, p 425

Redel DA, Hansmann M (1984) Fetale Echokardiographie – ihre Anwendung in Diagnostik und Therapie. Gynäkologe 17:41
Redel DA, Hansmann M, Dieberg S (1984) Pränatale Echokardiographie – Indikationen und Ergebnisse. In: Kowalewski S (Hrsg) 6. Symp Pädiatr Intensivmed. Thieme, Stuttgart New York
Sahn DJ, DeMario A, Kisslo J, Weyman A (1978) Recommendations regarding quantitations in M-mode echocardiography. Results of a survey of echocardiographic measurements. Circulation 58:1072
Sahn DJ, Lange LW, Allen HD, Goldberg SJ, Anderson C, Giles H, Haber K (1980) Quantitative real-time cross-sectional echocardiography in the developing normal human fetus and newborn. Circulation 62:588
Sahn DJ (1982) Two-dimensional echocardiographic method for identification of congenital heart malformations in unborn human fetuses. 1er symposium international d'echocardiologie foetale, Strasbourg 1982, p 119
Sahn DJ, Shenker L, Reed KL, Valdes-Cruz LM, Sobonya R, Anderson C (1982) Prenatal ultrasound diagnosis of hypoplastic left heart syndrome in utero associated with hydrops fetalis. Am Heart J 104:1368
Winter R, Müller WD, Beitzke A, Höfler H (1979) Pränatale Diagnose eines Herzfehlers mit Ultraschall. Z Geburtshilfe Perinatol 183:465
Wladimiroff JW (1981) Ultraschalluntersuchung des fetalen und neonatalen Herzens und des kardiovaskulären Systems. Ultraschall 2:221
Wladimiroff JW, McGhie J (1981) Ultrasonic assessment of cardiovascular geometry and function in the human fetus. Br J Obstet Gynaecol 88:870

Chapter 8

Bayer H, Issel EP, Schulte R (1972) Neue Meßgrößen bei der Erkennung einer intrauterinen Retardierung der Furcht mittels Ultraschalldiagnostik. Zentralbl Gynäkol 94:1169
Benson DM, Waldroup LD, Kurzt AB, Rose JL, Rifkin MD, Goldberg BB (1983) Ultrasonic tissue characterization of fetal lung, liver and placenta for the purpose of assessing fetal maturity. J Ultrasound Med 2:489
Bernaschek G, Dadak C, Kratochwil A (1980) Echographische Darstellung der großen fetalen Gefäße. Ultraschall 1:101
Bowie JD, Clair MR (1982) Fetal swallowing and regurgitation: observation of normal and abnormal activity. Radiology 144:877
Campbell S, Wilkin D (1975) Ultrasonic measurement of fetal abdomen circumference in the estimation of fetal weight. Br J Obstet Gynaecol 82:689
Chinn DH, Filly RA, Callen PW (1982) Ultrasonic evaluation of fetal umbilical and hepatic vascular anatomy. Radiology 144:153
Garrett WJ, Robinson DE (1971) Assessment of fetal size and growth by ultrasonic echoscopy. Obstet Gynaecol 38:525
Hansmann M, Voigt U (1973) Ultrasonic fetal thoracometry: an additional parameter for determining fetal growth. Excerpta Medica (Abstr), 2nd World Congress on Ultrasonics in Medicine, Rotterdam
Hansmann M (1975) Ultraschallkephalo- und Thorakometrie zur Kontrolle des fetalen Wachstums unter besonderer Berücksichtigung der praepartalen Gewichtsschätzung. Habilitationsschrift, Med Fakultät Bonn
Hansmann M, Hackelöer BJ, Staudach A (1985) Ultraschalldiagnostik in Geburtshilfe und Gynäkologie. Springer, Berlin Heidelberg New York Tokyo
Hansmann M, Hackelöer BJ, Staudach A (1986) Ultrasound diagnosis in obstetrics and gynecology. Springer, Berlin Heidelberg New York Tokyo
Higginbottom J, Slater J, Porter G (1975) Estimation of fetal weight from ultrasonic measurement of trunk circumference. Br J Obstet Gynaecol 82:698
Holländer HJ (1972, 1975, 1984) Die Ultraschalldiagnostik in der Schwangerschaft. Urban & Schwarzenberg, München Berlin Wien
Jeanty P, Romero R (1984) Obstetrical ultrasound. McGraw-Hill, New York
Jeanty P, Romero R, Hobbins JC (1984) Vascular anatomy of the fetus. J Ultrasound Med 3:113
Kossoff G (1981) New clinical applications. In: Kurjak A, Kratochwil A (eds): Recent advances in ultrasound diagnosis 3. Excerpta Medica International Congress Series 553

Kugener H, Hansmann M (1976) Zur Topographie einer Referenzebene für die Ultraschallthorakometrie. Z Geburtshilfe Perinatol 180:313
Lewi S, Erbsman F (1975) Antenatal fetal growth from the nineteenth week (Ultrasonic study of 12 head and chest dimensions). Am J Obstet Gynecol 121:262
Moore KL (1977) The developing human. Clinically oriented embryology, 2nd edn, 1977. Courtesy WB Saunders Co
Morin FR, Winsberg F (1978) Ultrasonic and radiographic study of the vessels of the fetal liver. J Clin Ultrasound 6:409
Schillinger H, Müller R, Kretzschmar M, Wode J (1975) Gewichtsbestimmung des Feten durch Ultraschall. Geburtshilfe Frauenheilkd 35:866
Schlensker KH, Decker I (1973) Voraussage des kindlichen Geburtsgewichtes aufgrund der Ultraschallkephalometrie und Thorakometrie am Feten. Geburtshilfe Frauenheilkd 33:859
Schmidt W, Yarkoni S, Jeanty P, Grannum P, Hobbins JC (1985) Sonographic measurements of the fetal spleen: Clinical implications. J Ultrasound Med 4:667
Thompson HE, Holmes JH, Gottesfeld KR, Taylor ES (1965) Fetal development as determined by ultrasonic pulse echo techniques. Am J Obstet Gynecol 92:44
Vandenberghe K, DeWolf F (1980) Ultrasonic assessment of fetal stomach function. Physiology and clinic. In: Kurjak A (ed): Recent advances in ultrasound diagnosis 2. Excerpta Medica International Congress Series 498
Wladimiroff JW, Leijs R, Smit B (1980) Human fetal stomach profiles. In: Kurjak A (ed): Recent advances in ultrasound diagnosis 2. Excerpta Medica International Congress Series 498

Chapter 9

Bernascheck G, Kratochwil A (1980) Echographische Studie über das Wachstum der fetalen Niere in der zweiten Schwangerschaftshälfte. Geburtshilfe Frauenheilkd 40:1059
Birnholz JC (1983) Determination of fetal sex. N Engl J Med 309:942
Bowie JD, Rosenberg ER, Andreotti RF, Fields SJ (1983) The changing sonographic appearance of fetal kidneys during pregnancy. J Ultrasound Med 2:505
Brusis E, Nitsch B, Wengeler H (1975) Fruchtwasser und Amnion. In: Wulf KH (Hrsg) Klinik der Frauenheilkunde und Geburtshilfe Bd VI Erg 1975, S 668. Urban & Schwarzenberg, München Wien Baltimore
Campbell S, Wladimiroff JW, Dewhurst CJ (1973) The antenatal measurement of fetal urine production. Br J Obstet Gynaecol 80:680
Cooper C, Mahony BS, Bowie JD, Pope II (1985) Prenatal ultrasound diagnosis of ambiguous genitalia. J Ultrasound Med 4:433
Deutinger J, Spernol R, Bernaschek G (1984) Können fetale Nierenbeckenerweiterungen physiologisch sein? Geburtshilfe Frauenheilkd 44:441
Eleyalde BR, DeEleyalde MM (1984) Further comments on amniocentesis in twin gestations. Am J Med Genet 17:699
Eleyalde BR, DeEleyalde MM, Heitman T (1985) Visualization of the fetal genitalia by ultrasonography: A review of the literature and analysis of its accuracy and ethical implications. J Ultrasound Med 4:633
Grannum P, Bracken M, Silverman R, Hobbins JC (1980) Assessment of kidney size in normal gestation by comparison of ratio of kidney circumference to abdominal circumference. Am J Obstet Gynecol 136:249
Hansmann M, Niesen H, Födisch HJ (1979) Pränatale Ultraschalldiagnose des Potter-Syndroms. Gynäkologe 12:69
Hansmann M (1984) Möglichkeiten und Grenzen sonographischer Diagnostik fetaler Erkrankungen und Mißbildungen. In: Kowalewski S (Hrsg) Pädiatrische Intensivmedizin VI. Thieme, Stuttgart New York S 56
Hansmann M, Hackelöer BJ, Staudach A (1985) Ultraschalldiagnostik in Geburtshilfe und Gynäkologie. Springer, Berlin Heidelberg New York Tokyo
Harrison MR (1983) Perinatal management of the fetus with a correctable defect. In: Callen PW (ed) Ultrasonography in obstetrics and gynecology. Saunders, Philadelphia London Toronto
Hoddick WK, Filly RA, Mahony BS, Callen PW (1985) Minimal fetal renal pyelectasis. J Ultrasound Med 4:85
Jeanty P, Romero R (1984) Obstetrical ultrasound. McGraw-Hill, New York
Kass M, Shaw MW (1976) The risk of birth defects and parents' right to know. Am J Law Med 2:213

Kenna TW (1973) The patient-physician relationship: present law and trends for the future implied in Cobbs vs. Grant. Univ San Francisco Law Review 8:320
Kratochwil A (1982) Sonographische Anatomie der normalen Schwangerschaft. Swiss Med 4:104
Kurjak A, Kirkinen P, Latin V, Ivankovic D (1981) Ultrasonic assessment of fetal kidney function in normal and complicated pregnancies. Am J Obstet Gynecol 141:266
Lewis E, Kurtz AB, Dubbins PA (1982) Realtime ultrasonographic evaluation of normal fetal adrenal glands. J Ultrasound Med 1:265
Maurer G, Winter R, Hofmann H, Müller WD, Ring E, Petritsch P (1985) Diagnostik und Therapie fetaler Nieren- und Harnwegsfehlbildungen. Ultraschall 6:173
McCrory WW (1972) Developmental nephrology. Cambridge, Harvard University Press p 40
Oliver J (1968) Nephrons and kidneys. Harper and Row, New York p 1
Porter KA (1978) The kidneys. In: Symmers (ed) Systemic pathology vol 4, p 1376. Churchill Livingstone, Edinburgh London New York
Potter EL (1972) Normal and abnormal development of the kidney. Yearbook Medical Publishers, Chicago
Schmidt W, Kubli F, Schroeder T (1981) Ultrasonographische Befunde beim „Potter-Syndrom". Geburtshilfe Frauenheilkd 41:374
Staudach A, Laßmann R, Rosenkranz W, Engels M, Joos H, Rücker J (1984) Praenatale Diagnose fetaler Entwicklungsstörungen – das Modell eines interdisziplinären Teams. In: Kowalewski S (Hrsg) Pädiatrische Intensivmedizin VI. Thieme, Stuttgart New York
Stenchever MA (1972) An abuse of prenatal diagnosis. JAMA 221:408
Stephens JD, Sherman S (1983) Letter to the editor. Determination of fetal sex by ultrasound. N Engl J Med 309:984
Stephens JD (1984) Prenatal diagnosis of testicular feminization. Lancet 2:1038
Visser GHA, Goodman JDS, Levine DH, Dawes GS (1981) Micturition and the heart period cycle in the human fetus. Br J Obstet Gynaecol 88:803
Weiß H, Zerres K, Hansmann M (1981) Pränatale Diagnose zystischer Nierenveränderungen mit Hilfe der Ultraschalltechnik. Ultraschall 2:244
Wladimiroff JW (1974) Effect of furosemide on fetal urine production. J Obst Gyn Brit Cwlth 82:221
Wladimiroff JW, Campbell S (1974) Fetal urine production rates in normal and complicated pregnancy. Lancet 1:151
Wladimiroff JW, Van Otterlo LC, Wallenburg HCS, Drogendijk AC (1976) A combined ultrasonic and biochemical study of fetal renal function in the term fetus. Eur J Obstet Gynecol Reprod Biol 6:103
Wladimiroff JW (1978) Studies of fetal physiology by sonography. In: DeVlieger M (ed) Handbook of clinical ultrasound. John Wiley & Sons, New York p 203
Zerres K (1981) Zystennieren, klinische, pathologisch anatomische und genetische Gesichtspunkte. Dissertation, Bonn
Zschoch H, Mahnke PF (1968) Die pathologische Anatomie des Kindesalters in der Sektionsstatistik. Fischer, Jena

Chapter 10

Bernaschek G (1982) Die Besonderheiten einer neuartigen echographischen Bestimmung der Kniegelenkskerne des Feten. Geburtshilfe Frauenheilkd 42:94
Chervenak FA, Tortora M, Hobbins JC (1985) Antenatal sonographic diagnosis of clubfoot. J Ultrasound Med 4:49
Filly RA, Golbus MS, Carey JC, Hall JG (1981) Short-limbed dwarfism: Ultrasonographic diagnosis by mensuration of fetal femoral length. Radiology 138:653
Hadlock FP, Harrist RB, Deter RL (1982) Ultrasonically measured fetal femur length as a predictor of menstrual age. Am J Radiol 138:875
Hadlock FP, Deter RL, Harrist RB (1983) A date-independent predictor of intrauterine growth retardation: femur length/abdominal circumference ratio. Am J Radiol 141:979
Hadlock FP, Harrist RB, Deter RL (1983) A prospective evaluation of fetal femur length as a predictor of gestational age. J Ultrasound Med 2:111
Hadlock FP,Harrist RB, Carpenter RJ (1984) Sonographic estimation of fetal weight: the value of femur length in addition to head and abdomen measurements. Radiology 150:535

Hansmann M, Hackelöer BJ, Staudach A (1985) Ultraschalldiagnostik in Geburtshilfe und Gynäkologie. Springer, Berlin Heidelberg New York Tokyo
Hills D, Buzzi K, Lawson W (1982) Off-axis dependence of sector scanner as a source of inherent error in measuring femur length. J Ultrasound Med 1 (suppl):101 (Abstract No 527)
Hobbins JC, Mahoney MJ (1980) The diagnosis of skeletal dysplasias with ultrasound. In: Sanders RC, James AE (eds) The principles and practice of ultrasonography in obstretics and gynecology, 2nd edn. Appleton-Century-Crofts, New York p 191
Hobbins JC, Bracken MB, Mahoney MJ (1982) Diagnosis of fetal skeletal dysplasias with ultrasound. Am J Obstet Gynecol 142:306
Hoffbauer H, Pachaly I, Arabin B (1978) Fetale Ultraschall-Somatometrie, Ultraschalldiagnostik. Thieme, Stuttgart
Hoffbauer H (1981) Sonographic measurement of fetal extremities. 4th European Congress on Ultrasonics in Medicine, Dubrovnik 1981
Hohler CW, Quetel TA (1981) Comparison of ultrasound femur length and biparietal diameter in late pregnancy. Am J Obstet Gynecol 141:759
Hohler CW. Quetel TA (1982) Fetal femur length: equations for computer calculation of gestational age from ultrasound measurements. Am J Obstet Gynecol 143:479
Holländer HF (1972) Die Ultraschalldiagnostik in der Schwangerschaft. Urban & Schwarzenberg, München Berlin Wien
Jeanty P, Rodesch F, Delbeke D (1984) Estimation of gestational age from measurements of fetal long bones. J Ultrasound Med 3:75
Jeanty P, Beck GJ, Chervenak FA, Kremkau FW, Hobbins JC (1985) A comparison of sector and linear array scanners for the measurement of the fetal femur. J Ultrasound Med 4:525
Jeanty P, Romero R, D'Alton M, Venus I, Hobbins JC (1985) In utero sonographic detection of hand and foot deformities. J Ultrasound Med 4:595
Lang M, Hansmann M, Bellmann O, Azubuike J (1979) Thanatophorer Zwergwuchs – pränatale Diagnostik und Geburtsleitung. Gynäkologe 12:84
Leo FP, Graham D, Cordier JM (1983) Length discrepancies produced by mechanical sector scanners. J Ultrasound Med 2 (suppl):193 (Abstract No 1730)
Leo FP, Sanders RC, Graham D (1983) Discrepancies in femur length due to the type of realtime ultrasound system used for the study. RSNA Abstract No 772, Scientific Program p 257
Luthy DA, Hall JG, Graham CB (1979) Prenatal diagnosis of thrombocytopenia with absent radii. Clin Genet 15:495
Mahoney MJ, Hobbins JC (1977) Prenatal diagnosis of chondroectodermal dysplasia (Ellis-vanCreveld syndrome) using fetoscopy and ultrasound. N Engl J Med 297:258
Mahony BS, Filly RA (1984) High-resolution sonography assessment of the fetal extremities. J Ultrasound Med 3:489
O'Brien GD, Queenan JT (1981) Growth of the ultrasound fetal femur length during normal pregnancy. Part I. Am J Obstet Gynecol 141:833
O'Brien GD, Queenan JT, Campbell S (1981) Assessment of gestational age in the second trimester by real-time ultrasound measurement of the femur length. Am J Obstet Gynecol 139:540
O'Brien GD, Queenan JT (1982) Ultrasound fetal femur length in relation to intrauterine growth retardation. Part II. Am J Obstet Gynecol 144:35
Pretorius D, Nelson T, Johnson ML (1983) Measurement errors and their impact on the evaluation of fetal age by ultrasound. RSNA Abstract No 138, Scientific Program p 49
Queenan JT, O'Brien GD, Campbell S (1980) Ultrasound measurement of fetal limb bones. Am J Obstet Gynecol 138:297
Richardson MM, Beaudet AL, Wagner ML, Malini S, Rosenberg HS, Lucci JA (1977) Prenatal diagnosis of recurrence of saldino-noonan dwarfism. J Pediat 91:467
Schlensker KH (1981) Die sonographische Darstellung der fetalen Extremitäten im mittleren Trimenon. Geburtshilfe Frauenheilkd 41:366
Schlensker KH (1982) Biometrie der fetalen Extremitäten. Swiss Med 4:140
Smith WL, Breitweiser TD, Dinno N (1981) In utero diagnosis of achondrogenesis typ I. Clin Genet 19:51
Smith DW (1982) Recognizable patterns of human malformations. Saunders, Philadelphia London Toronto
Staudach A, Laßmann R, Menzel C (1982) Mißbildungsdiagnostik vor der 24. Woche. In: Kratochwil A, Reinold E (Hrsg) Ultraschalldiagnostik 81. Thieme, Stuttgart New York.
Staudach A, Laßmann R (1984) Ultraschalldiagnostik von fetalen Mißbildungen. Oester Aerztetg 39/7:476

Staudach A, Laßmann R, Rosenkranz W, Engels M, Joos H, Rücker J (1984) Praenatale Diagnose fetaler Entwicklungsstörungen – das Modell eines interdisziplinären Teams. In: Kowalewski S (Hrsg) Pädiatrische Intensivmedizin VI. Thieme, Stuttgart New York

Terinde R, Driedger E, Koslowski P (1981) Ultrasound biometry of fetal extremities by measurement of fetal limb bones. 4th European Congress on Ultrasonics in Medicine, Dubrovnik 1981

Winsberg F (1983) Accuracy of measurements with linear and sector scanner (Letter to the Editor) J Clin Ultrasound 11:A10

Winter R, Rosenkranz W, Hofmann H, Zierler H, Becker H, Borkenstein M (1985) Prenatal diagnosis of campomelic dysplasia by ultrasonography. Prenatal Diagnosis 5:1

Yarkoni S, Schmitt W, Jeanty P, Reece EA, Hobbins JC (1985) Clavicular measurement: a new biometric parameter for fetal evaluation. J Ultrasound Med 4:467

Yeh MN, Barrow B, Braceroh (1981) Ultrasound measurement of the femur length as an index of fetal growth and development. In: Proceedings of the 26th Annual Meeting of the AIUM. Bethesda, American Institute of Ultrasound in Medicine, 1981 p 35

Subject Index

MIX
Papier aus verantwortungsvollen Quellen
Paper from responsible sources
FSC® C105338

If you have any concerns about our products,
you can contact us on
ProductSafety@springernature.com

In case Publisher is established outside the EU,
the EU authorized representative is:
Springer Nature Customer Service Center GmbH
Europaplatz 3, 69115 Heidelberg, Germany

Printed by Libri Plureos GmbH
in Hamburg, Germany